FREDERICK SODDY (1877–1956)

CHEMISTS AND CHEMISTRY

*A series of books devoted to the examination of the history and development of chemistry from its early emergence as a separate discipline to the present day. The series will describe the personalities, processes, and theoretical and technical advances which have shaped our current understanding of chemical science.*

1877
1956
FREDERICK SODDY
F.R.S. NOBEL LAUREATE
LECTURER IN PHYSICAL CHEMISTRY (INCLUDING RADIOACTIVITY) GLASGOW UNIVERSITY, 1904–1914
WITH RUTHERFORD HE SHARED THE DISCOVERY OF ATOMIC DISINTEGRATION, WITH RAMSAY THE PROOF OF HELIUM PRODUCTION FROM α-PARTICLES. IN THIS DEPARTMENT HE FORMULATED HIS CONCEPTION OF ISOTOPES

# FREDERICK SODDY (1877-1956)

*Early Pioneer in Radiochemistry*

*Edited by*

GEORGE B. KAUFFMAN

*Professor of Chemistry*
*California State University, Fresno, CA 93740, U.S.A.*

D. REIDEL PUBLISHING COMPANY

A MEMBER OF THE KLUWER ACADEMIC PUBLISHERS GROUP

DORDRECHT / BOSTON / LANCASTER / TOKYO

Library of Congress Cataloging in Publication Data

CIP-data appear on separate card
ISBN 90-277-1926-8
LC CARD NUMBER 85-28115

Published by D. Reidel Publishing Company.
P.O. Box 17, 3300 AA Dordrecht, Holland.

Sold and distributed in the U.S.A. and Canada
by Kluwer Academic Publishers,
190 Old Derby Street, Hingham, MA 02043, U.S.A.

In all other countries, sold and distributed
by Kluwer Academic Publishers Group,
P.O. Box 322, 3300 AH Dordrecht, Holland.



Printed in The Netherlands.

*To*
*my wife,*
*Laurie*

'Grow old along with me!
The best is yet to be,
The last of life, for which the first was made.'

Robert Browning
*Rabbi Ben Ezra*

'Die Geschichte der Wissenschaft ist die Wissenschaft selbst.'
(The history of science is science itself.)

Johann Wolfgang von Goethe
*Mineralogie und Geologie*

## TABLE OF CONTENTS

# EDITOR'S PREFACE

On August 18, 1977 a special 'Soddy Session' was held at the Fifteenth International Congress of the History of Science, Edinburgh, Scotland, with Dr. Thaddeus J. Trenn as Symposium Chairman. This session was organized to commemorate the 100th anniversary of the birth of Frederick Soddy (born September 2, 1877, Eastbourne, England; died September 22, 1956, Brighton, England), who was awarded the 1921 Nobel Prize in Chemistry 'for his contributions to our knowledge of the chemistry of radioactive substances, and his investigations into the origin and nature of isotopes'. Soddy taught and/or carried out research at Oxford University (where he was Lee's Professor of Chemistry), McGill University (where he and Sir Ernest Rutherford proposed the disintegration theory of radioactivity), University College, London (where he and Sir William Ramsay demonstrated natural transmutation), Glasgow University (where he formulated his displacement law and concept of isotopes), and Aberdeen University. In addition to his contributions to radiochemistry, he proposed a number of controversial economic, social, and political theories.

The present volume contains the eight lectures presented at the symposium, two additional papers written especially for this volume (Kauffman, Chapter 4 and Krivomazov, Chapter 6), a paper on Soddy's economic thought (Daly, Chapter 11), and three selections from Soddy's works. Furthermore, an introductory account of Soddy's life and work by Thaddeus J. Trenn as well as a Soddy chronology, and name and subject indexes compiled by the editor are provided. Since the essays are written by internationally prominent authorities, it is hoped that this volume will be of interest and use to historians of science, scientists, and students of radioactivity and atomic structure as well.

In editing the papers that comprise this volume, I have kept changes to an absolute minimum. I have not Americanized the speling of British authors. On several papers by authors whose native languages are not English (van Spronsen, Chapter 5; Krivomazov, Chapter 6; and Tanaka and Yamasaki, Chapter 7), I have taken the liberty of rendering the texts into correct English. In a number of cases, I have also provided abstracts and references. I have made no attempt to eliminate duplication inasmuch as repetitions of the same or similar material by different authors have presented different viewpoints and emphases of the work of an extremely multifaceted scientist.

For their kind permission to reprint several of the papers in this volume I wish to thank Charles Scribner's Sons, New York and the British Association for the Advancement of Science ('Frederick Soddy' by Thaddeus J. Trenn), *Scientia* (Soddy's 'Transmutation, the Vital Problem of the Future'), Macmillan Journals, Ltd. (Soddy's 'Intra-atomic Charge'), Allen and Unwin, Ltd., London (Foreword to Soddy's 'The Frustration of Science'), the *British Journal for the History of Science* (Badash, Chapter 1; Cruickshank, Chapter 8; Freedman, Chapter 9; and Trenn, Chapter 10), and Duke University Press (Daly, Chapter 11). I also wish to thank the Chemical Society for permission to reproduce the Soddy plaque from Glasgow University (*Proc. Chem. Soc.*, (1963): facing p. 328).

It is my pleasure to acknowledge the generous financial assistance of the History and Philosophy of Science Program of the National Science Foundation Division of Social Sciences (Grant SOC76-11267). I am also indebted to Thaddeus J. Trenn and Lawrence Badash for valuable suggestions, to the California State University, Fresno for a sabbatical leave, to Julie A. Hildebrand, Marlene Frye, Donna Hamm, and Mary McIntyre for typing the manuscript, and to Robert Michelotti for the illustrations. Last but not least, I wish to thank my wife Laurie for her support and encouragement.

GEORGE B. KAUFFMAN

*Cabin C8*
*Lakeshore Village*
*Lakeshore, CA 93634, U.S.A.*

# INTRODUCTION TO: FREDERICK SODDY (1877–1956) EARLY PIONEER IN RADIOCHEMISTRY

Credit for the important discoveries in the natural radioactivity of the heavy elements that took place near the end of the last century and extended into the second decade of this century should be divided almost equally between chemists and physicists. Their pioneering work laid the foundation for our understanding of the structure of the atom with its central nucleus, which in turn was basic to our subsequent understanding of nuclear structure, molecular structure, and our present general understanding of both inanimate and living matter.

Frederick Soddy was one of the trail blazers among the chemists, the practitioners in the field of radiochemistry, a group which included such other giants as Marie Curie, Otto Hahn, Kasimir Fajans, Fritz Paneth, and George Hevesy. Material on Soddy and those remarkable early days of the investigations of natural radioactivity is always interesting and valuable. It is therefore fortunate that George B. Kauffman, winner of the 1978 Dexter Award in the History of Chemistry, has made available in this book eight essays based on the talks given at the Fifteenth International Congress of the History of Science, held in 1977 to commemorate the 100th anniversary of Soddy's birth. The essays are written by internationally recognized authorities with intimate knowledge of the subject, including some who had personal acquaintanceships with Soddy, and they are augmented by the inclusion of additional authoritative essays and some of Soddy's most important original publications.

Although Soddy is generally credited with playing a seminal role in leading us to our understanding of the concept of isotopes, the importance of his intuition and interpretative analysis of experiments in leading to the understanding of radioactivity is not so well known. Some months after his arrival at McGill University in Montreal, Canada, he began his collaborative research, extending over a period of nearly two years (from 1901 to 1903) with Ernest Rutherford, who was just beginning his remarkable career of research accomplishments that led to his rating as the leading historical investigator in the field of nuclear science. After a series of investigations, with some confusing and con-

*George B. Kauffman (ed.), Frederick Soddy (1877–1956)*, xiii–xiv.

tradictory results, by Rutherford and Soddy and investigators in other laboratories on the chemical and radioactive properties of the decay products of uranium, thorium, and radium (some of which are gaseous), Rutherford and Soddy published in 1902 a beautiful, lengthy account in which they suggested that radioactivity is accompanied by a chemical change or 'transmutation' of the elements. Studies of the nature of the emitted radiation and the behavior of the decay products continued and finally culminated in Rutherford and Soddy's paper of 1903, in which they summarized all these early investigations of the nature of the radioactive decay process and made some interesting estimates of the energy release in radioactive decay. In this connection, a characterization by Norman Feather, an eminent British nuclear physicist and acquaintance of Soddy's during his later years and a contributor to this book, is very illuminating: 'It must be conceded that the Fates were kind to Rutherford. He was left in Canada to discover that his collaboration with a young Oxford chemist, Frederick Soddy, was to mean more to him at that precious juncture than any Chair in Europe'.

The final resolution of what was happening in radioactive decay was described in a series of papers by Soddy and by Fajans. In these papers the new radioactive species were given places in the Periodic Table of the Elements. As a result Soddy clearly recognized the need for the concept of isotope, which gave several radioactive species the same place in the Periodic Table, but still differentiated them by virtue of their different masses.

Soddy turned away from research in radioactivity and related fields rather early in his life, as illustrated by the later papers in this volume. He became involved with the problems of energy sources, economics, and sociology. He was driven to this by his perception of the need to work for the improvement of the lot of mankind and his feeling that efforts must be made to direct the fruits of science to beneficent uses in order to help achieve this goal. He expressed this well in the following quotation from his writings: 'It is modern science which has made the modern world great, with a greatness that the illustrious epochs of history cannot match'.

*January, 1984*

GLENN T. SEABORG

*Lawrence Berkeley Laboratory*
*Berkeley, California*

## A SODDY CHRONOLOGY

| *Date* | *Age* | |
|---|---|---|
| 1877 | | Born Sept. 2 at Eastbourne, Sussex, England |
| 1892–1894 | 15–17 | Attends Eastbourne College |
| 1894–1895 | 17–18 | Attends University College of Wales, Aberystwyth to study classics |
| 1895 | 18 | Awarded science scholarship to Merton College, Oxford |
| 1896–1898 | 18–20 | Studies chemistry at Merton College and graduates with first-class honors B.A. |
| 1898–1900 | 20–21 | Independent chemical research at Oxford |
| 1899 | 21 | Fellow of the Chemical Society |
| 1900–1902 | 23–25 | Demonstrator in Chemistry Department, McGill University, Montreal, Canada |
| 1901 (mid-Oct.)–1903 (mid-Apr.) | 24–25 | 18-month collaboration with Ernest Rutherford resulting in nine papers |
| 1902 (Apr.) | 24 | Discovery of thorium-X and proposal of Transformation Theory (E. Rutherford and F. Soddy *J. Chem. Soc.* **81** (1902): 837–60) |
| 1903 (Apr.) | 25 | Disintegration Theory of Radioactivity (E. Rutherford and F. Soddy *Phil. Mag.* **5** (1903): 576–91) |
| 1903–1904 | 26 | Work at University College, London with William Ramsay; continues study of radium emanation; demonstrates spectroscopically the production of helium from radium as predicted by Rutherford and Soddy – the first clear experimental proof of the natural transmutation of elements (W. Ramsay and F. Soddy *Nature* **68** (August 13, 1903): 354) |
| 1904 | 27 | *Radio-Activity* (book); demonstrates that |

| | | |
|---|---|---|
| | | uranium is not directly transformed into radium (*Nature* **70** (May 12, 1904): 30) |
| 1904–1914 | 26–36 | Lecturer in Physical Chemistry and Radioactivity, Glasgow University, Scotland; decade of research on radiochemistry, nonseparable radioelements, and relationship between radioelements and periodic table |
| 1904–1920 | 26–42 | Contributes series of original reports on 'Radioactivity' for *Annual Reports on the Progress of Chemistry* |
| 1908 | 30 | *The Interpretation of Radium* (book) |
| 1908–1936 | 30–59 | Marriage to Winifred Beilby, terminated by wife's death |
| 1909 | 31 | Extension of Disintegration Theory to branching series (*Phil. Mag.* **18** (1909): 739–44) |
| 1910 | 33 | Elected Fellow of Royal Society; elucidation of chemistry of mesothorium I; advances the concept (not the term) of isotopy (*Ann. Reports* **7** (1910): 286) |
| 1911 | 33 | Proposes alpha-ray rule (*The Chemistry of the Radio-Elements*, *Part I*, p. 29) |
| 1912 | 34 | *Matter and Energy* (book) |
| 1913 (Feb.) | 35–36 | Enunciates the Displacement Law; separation of actinium series (*Brit. Ass. Rep.* (1913): 446); awarded Cannizzaro prize; introduces the term 'isotope' (*Nature* **92** (December 4, 1913): 400) |
| 1914 | 36 | *The Chemistry of the Radio-Elements*, *Part II* (book containing new views on isotopes and periodic law); determines atomic weight of lead from thorite demonstrating that common lead is a mixture of isotopes (F. Soddy and H. Hyman *Proc. Chem. Soc.* **30** (1914): 134–5) |
| 1914–1919 | 37–41 | Professor of Chemistry, Aberdeen University, Scotland; research hampered by war work |

| | | |
|---|---|---|
| 1918 | 40 | Discovers the direct parent of actinium, isotopic with U-$X_2$ and later called protactinium (F. Soddy and J. A. Cranston *Proc. Roy. Soc.* **94A** (1918): 384; *Nature* **100** (1918): 498–9) |
| 1919–1936 | 42–59 | Lee's Professor of Chemistry, Oxford University; interest shifts to social implications of science; becomes critic of economics, business, finance, and sociology |
| 1920 | 42 | *Science and Life* (book) |
| 1922 | 44 | Awarded 1921 Nobel Prize in Chemistry 'for his contributions to our knowledge of the chemistry of radioactive substances, and his investigations into the origin and nature of isotopes' |
| 1922 | 46 | *Cartesian Economics* (booklet) |
| 1923 | 47 | Process for extraction of thorium |
| 1924 | 48 | *The Inversion of Science and a Scheme of Scientific Reformation* (book) |
| 1926 | 50 | *Wealth, Virtual Wealth and Debt* (book) |
| 1931 | 54 | *Money vs. Man* (book) |
| 1932 | 55 | *The Interpretation of the Atom* (book) |
| 1934 | 57 | *The Role of Money* (book) |
| 1935 | 58 | *The Frustration of Science* (book edited by Sir Daniel Hall whose foreword by Soddy is a classic call for responsibility in science) |
| 1936 | 59 | Early retirement from Oxford University after wife's death |
| 1940 | 63 | Patents 1923 process for thorium extraction |
| 1943 | 66 | *The Arch-enemy of Economic Freedom* (booklet) |
| 1949 | 72 | *The Story of Atomic Energy* (book) |
| 1951 | 74 | Awarded the Albert Medal |
| 1953 | 76 | *Atomic Transmutation* (book) |
| 1954 | 77 | *Isotopy* (booklet co-authored by J. A. Cranston) |
| 1956 | 79 | Death at Brighton, East Sussex, England, September 22 |

THADDEUS J. TRENN

# FREDERICK SODDY*

SODDY, FREDERICK (*b.* Eastbourne, England, 2 September 1877: *d.* Brighton, England, 22 September 1956), *radiochemistry*, *science and society*.

Soddy developed with Lord Rutherford during 1901–1903 the disintegration theory of radioactivity, confirmed with Sir William Ramsay in 1903 the production of helium from radium, advanced in 1910 the *concept* of isotope, proposed in 1911 the alpha-ray rule leading to the full displacement law of 1913, and was the 1921 Nobel laureate in chemistry, principally for his investigations into the origin and nature of isotopes.

The youngest son of a London merchant, Soddy was raised in the Calvinist tradition by his dominant half-sister. He developed a lifelong sense of extreme social independence, as well as a plague-on-both-your-houses attitude toward religious controversy, later extended to social institutions in general. An aspiring scientist from an early age, Soddy was encouraged by his influential science master, R. E. Hughes, at Eastbourne College to study chemistry at Oxford. After an interim year at co-educational Aberystwyth, Soddy in 1895 received a science scholarship to Merton College, Oxford. In 1898, with Ramsay as external examiner, Soddy received a first-class honors degree; he remained at Oxford for two more years, engaged in independent chemical research.

In May 1900, Soddy adventurously followed up an unsuccessful application to Toronto with a personal visit to Montreal, accepting a position as demonstrator at McGill University. His childless marriage in 1908 to Winifred Beilby (*d.* 1936) was a source of great happiness and stability in his life. Soddy was 'an admirable writer and a clear and interesting lecturer,'[1] noted for originality in demonstrations. A fellow of both the Chemical Society (from 1899) and the Royal Society (from 1910), Soddy was also a foreign member of the Swedish, Italian, and Russian academies of science. In 1913 he was awarded the Cannizzaro Prize for his important contributions to the new chemistry.

Profoundly disturbed by World War I and 'enraged'[2] by the death of

*George B. Kauffman (ed.), Frederick Soddy (1877–1956)*, xix–xxviii.

Moseley, Soddy felt that society was not yet sufficiently mature to handle properly the advances of science. He began to concern himself more with the interaction between science and society. In order to ascertain 'why so far the progress of science has proved as much a curse as a blessing to humanity,'[3] Soddy studied economics. He considered the free development of science to be the new wealth of nations and advocated the rise of a 'scientific civilization.'[4] The 'curse' he felt arose from constraints, put upon both the progress of science and the distribution of technological productivity, for the self-maintenance of the existing but decadent economic system.[5]

At McGill, Soddy joined with Rutherford in a series of investigations which produced the theoretical explanation of radioactivity. The constant production of a material 'emanation' from thorium was shown to be the combined effect of the production of an intermediate, but chemically separable, substance, thorium X, balanced by its decay. The production of one substance was thus the result of the uncontrollable disintegration of another. The 'radiation' proved to be both particulate in nature and a direct accompaniment of the process of disintegration. The rate of the process was found in every case to be as the exponential law of a monomolecular chemical reaction. So complete was their 1903 disintegration theory of radioactivity, that in 1909 only an extension to branching series was required.[6]

In March 1903, Soddy elected to join Ramsay in London to examine more fully the gaseous products of decay. Using Giesel's radium preparations, Ramsay and Soddy experimentally confirmed in July 1903 the prediction of Rutherford and Soddy that radium would continuously produce helium. In 1908 Rutherford 'settled for good'[7] the long-suspected identity of the helium, so produced, with expelled alpha particles. During the ten-year period following his 1904 appointment to the University of Glasgow, Soddy helped to clarify the relation between the plethora of radioelements and the periodic table.

McCoy and Ross had reported in 1907 that Hahn's 1905 radiothorium was chemically inseparable from thorium. Boltwood, in turn, indicated a similar difficulty with thorium and ionium. From crystal morphology studies, Strömholm and Svedberg in 1909 confirmed a family resemblance between such radioelements as thorium X and radium. In 1910 the chemical inseparability of mesothorium I and radium, reported by Marckwald, as well as Soddy's own experimental evidence, that these two radioelements form an inseparable trio with thorium X, convinced

Soddy that such cases of chemical inseparability were actually chemical *identities*. Without the unnecessary continuation of the genetic series of radioelements throughout the entire periodic table, postulated by Strömholm and Svedberg, Soddy declared in 1910 that 'the recognition that elements of different atomic weight may possess identical chemical properties seems destined to have its most important application in the region of the inactive elements.'[8] 'Soddy possessed,' as Hahn wrote in admiration,[9] 'the courage to declare that these were chemically identical elements.'

To be able to refer generically to these active and inactive elements with identical chemical properties, Soddy introduced the technical term 'isotope' in 1913.[10] While chemically inseparable, active isotopes were distinguishable by their radioactive properties, and all isotopes differed in atomic weight. Soddy suggested that the 1912 metaneon of J. J. Thomson be considered 'a case of isotopic elements outside the radioactive sequences.'[11] Following Soddy, Aston announced a partial separation in 1913 on this very basis.[12] The connection between chemical properties and the periodic table became increasingly clarified with concurrent developments in the physics and chemistry of the nuclear atom. From the chemical side, Soddy proposed the alpha-ray rule in 1911, the key to the first of two locks. Applying his general principle that the common elements are mixtures of chemically inseparable elements 'differing step-wise by whole units of atomic weight'[13] specifically to the case of the radioelements, Soddy recognized that the expulsion of an alpha particle would result in a lighter element chemically inseparable from those occupying the 'next but one'[14] position in the periodic table. The second lock to the displacement law involved the beta transitions.

During 1912 Soddy assigned Lord Fleck the task of sorting out the short-lived beta emitters, especially at the complex branching points in the series. Once these experimental results became available, several partially correct generalizations were published, inducing Soddy, therefore, to publish his own complete and correct form of the law in February 1913. 'Fajans,' Soddy acknowledged, 'worked out the Periodic Law Generalization quite independently of me,'[15] although his conclusions were fundamentally different. On electrochemical considerations, Fajans interpreted the changes among the clusters, 'plejade,'[16] of radioelements as evidence against the nuclear origin of radio-changes.[17] Soddy, on the other hand, argued for a crucial distinction[18] between radio-change and chemical change, concluding on chemical evidence, as Bohr had done on physical evidence, that beta decay, like alpha decay, was of nuclear origin. As a result, Soddy considered van den Broek's hypothesis, that successive places in the periodic table correspond to unit differences in the net intra-atomic charge (see Figure 1) 'practically proved so far as the ... end of the sequence, from thallium to uranium, is concerned.'[19]

By early 1914 Moseley,[20] using physical methods, had completed his independent extension of this verification throughout the entire periodic table. During the period 1914–1919, in the chair of chemistry at Aberdeen, Soddy, in addition to his war work, examined two predictions of the displacement law. It was commonly accepted that lead was the end product only of the uranium series, and Soddy had predicted by 1913 that a heavier isotope of lead from thorium must also exist. Separate determinations were undertaken in 1914 on lead from Ceylon thorite by Soddy and on lead from uranium ores by T. W. Richards and O. Hönigschmid, thereby confirming the prediction that common lead was indeed a mixture of isotopes.[21] Soddy suggested that the parent of actinium might be an alpha-decaying member of Mendeleev's missing eka-tantalum. An exclusively beta-emitting homologue of tantalum found by Fajans and O. Göhring in 1913 and called 'brevium' [$UrX_2$], however, caused Soddy to begin to investigate the other alternative. But after proving that the parent of actinium could not be a beta-decay product of radium, he reexamined the first alternative with Cranston. In 1918 they found, isotopic with $UrX_2$, the direct parent of actinium, produced through the rare UrY branch, which was found in 1911 by G. Antonoff and later linked to uranium 235. Protactinium, element 91, was simultaneously and independently found by Hahn and Meitner.

Soddy was called in 1919 to a chair of chemistry at Oxford. During his seventeen-year tenure, he failed to establish the expected school of radiochemistry, devoting himself rather to the improvement of chemistry teaching and to the modernization of the laboratories.[22] He also continued to treat radioactive minerals for their constituents. After the disturbing death of his wife, Soddy retired early. He went exploring for monazite sand, and patented his 1923 process for thorium extraction in 1940. He then turned his attention to mathematics. Looking beyond to the significance of science, Soddy, who had once confidently spoken of the potential peaceful benefits for society given the key to 'unlock this

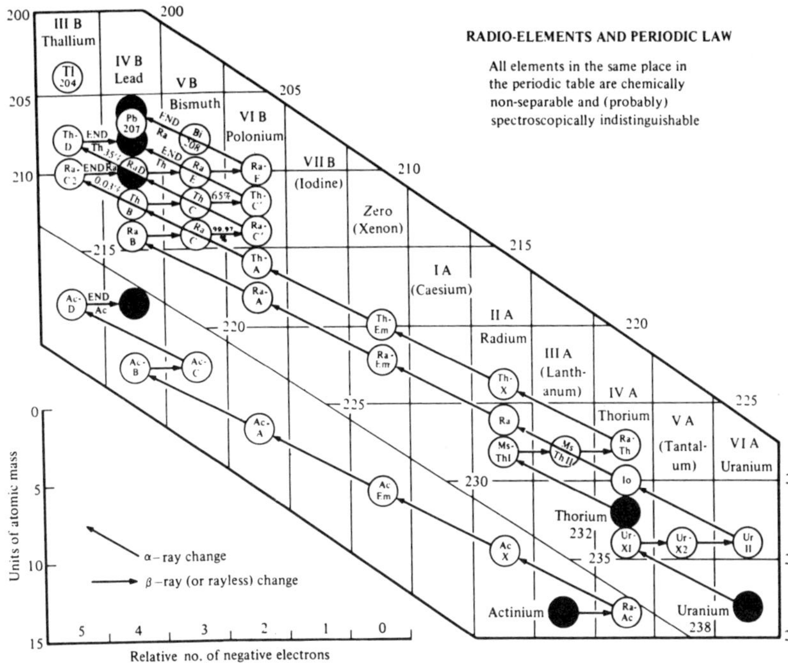


Fig. 1. Reproduced with the permission of the British Association for the Advancement of Science.

great store of energy bound up in the structure of the element,'[23] and, by controlling it, 'virtually provide anyone who wanted it with a private sun of his own.'[24] was profoundly concerned by subsequent developments. He zealously endeavored[25] to awaken the conscience of the scientific community to the social relevance of their own research. Soddy urged that 'universities and learned societies should no longer evade their responsibilities and hide under the guise of false humility as the hired servants of the world their work has made possible, but do that for which they are supported in cultured release from routine occupations, and speak the truth though the heavens fall.'[26] He was largely unheeded, however, and he judged at the end that the blame for the plight of civilization 'must rest on scientific men, equally with others, for being incapable of accepting the responsibility for the profound social upheavals which their own work primarily has brought about in human relationships.'[27]

THADDEUS J. TRENN

*RR4 – Box 639*
*Colborne Ontario*
*KOK 150 Canada*

## Notes

* Reprinted from Gillispie, Charles Coulston (Editor) *Dictionary of Scientific Biography*. New York: Charles Scribner's Sons, 1975, Vol. **12**, pp. 504–9.

[1] Rutherford letter, 15 June 1914, Cf. Howorth, *Pioneer Research*, 192. The original is in Bodleian Library, Soddy-Howorth Collection, 75, 95, courtesy Soddy trustees.

[2] Soddy, *Memoirs* **1**, 274.

[3] Fleck, 'Soddy', 210, courtesy the Royal Society.

[4] Soddy, 'Social Relations of Science', in *Nature* **141** (1938), 784–785.

[5] A comprehensive statement of his general view regarding the monetary system preventing modern Western civilization from distributing its scientific and technological abundance by peaceful means appears in an address, February 1950, partially republished in the 24-page *Commemoration to Professor Frederick Soddy* (London, 1958).

[6] Soddy, 'Multiple Atomic Disintegration: A Suggestion in Radioactive Theory', in *Philosophical Magazine* **18** (1909), 739–744; this was developed in 'Multiple Disintegration', in *Annual Report* **9** (1912), 311–316.

[7] Referring to his joint paper with Royds, *Philosophical Magazine* **17** (1909), 281. Rutherford further noted in his letter of 14 Feb. 1909 to Elster and Geitel that 'you will have seen that the $\alpha$ particle has at last been proved to be helium.' Darmstaedter Collection, G 1, 1896 (26), courtesy Staatsbibliothek, Preussischer Kulturbesitz.

[8] Soddy, 'Radioactivity', in *Annual Report* **7** (1910), 286. Strikingly similar views regarding mixtures of similar elements of different atomic weight were expressed by D. Strömholm and T. Svedberg in *Zeitschrift für Anorganische Chemie* **63** (1909), 206.

[9] Fleck, 'Soddy', 208. The rare earths had given ample evidence of chemical 'inseparability' without identity.

[10] Soddy, 'Intra-atomic Charge', in *Nature* **92** (4 Dec. 1913), 400. 'The same algebraic sum of the positive and negative charges in the nucleus, when the arithmetic sum is different, gives what I call "isotopes" ... because they occupy the same [*iso*] place [*topos*] in the periodic table' (see diagram). Perhaps the first use of 'isotope' for the position of elements was W. Preyer, *Das Genetische System der chemischen Elemente* (Berlin, 1893). The stimulus for Soddy's term arose when he 'got tired of writing "elements chemically identical and non-separable by chemical methods" and coined the name *isotope* ...,' as he said in 'Contribution to a Discussion on Isotopes,' in *Proceedings of the Royal Society* **99** (1921), 98.

[11] Soddy, 'Radioactivity', in *Annual Report* **10** (1913), 265.
[12] Aston, *Isotopes* **37**, 42.
[13] Soddy, 'The Chemistry of Mesothorium', in *Transactions of the Chemical Society* **99** (1911), 82: cf. n. 8.
[14] Soddy, *The Chemistry of the Radioelements* (1911), 29. For a remarkable partial anticipation of isotopes and the displacement law, see A. T. Cameron, *Radiochemistry* (London, 1910), 141.
[15] Soddy letter to F. O. Giesel, *ca.* 1913/14 in Giesel Archives, courtesy Chininfabrik, Buchler & Co., Brunswick. The generalization of A. S. Russell had not only assumed a *discontinuous* series, *Chemical News* **107** (31 Jan. 1913), 49, but also questioned the chemical identity notion of Soddy; cf. Russell letter to Rutherford, 14 Sept. 1912, Cambridge Univ. Lib., Add. MSS 7653/R106. Russell 'knew of Fleck's results', and 'through him they got known to Hevesy and Fajans'; cf. *Report of the British Association for the Advancement of Science* (1913), 446; and the Soddy letter to Howorth, 29 Jan. 1953, Bodleian Lib., Soddy Collection, Alton 29, item no. Trenn S–6.
[16] K. Fajans, *Radioaktivität und die neueste Entwickelung der Lehre von den chemischen Elementen* (Brunswick, 1919), 35.
[17] Fajans' letter to Rutherford, 10 April 1913, Cambridge Univ. Lib., Add. MSS 7653/F5.
[18] Soddy's distinction between chemical change and radio-change was orginally based upon the disintegration theory, 'Radioactive Change', in *Philosophical Magazine* **5** (1903), 576. With the development of the nuclear atom, however, it became possible to clarify this distinction by defining the actual locus of the radio-changes. Bohr expressed this clarification in his letter to Hevesy, 7 Feb. 1913, L. Rosenfeld, 'Introduction' to *On the Constitution of Atoms and Molecules* (Copenhagen, 1963), xxxii.
[19] Soddy, 'Intra-atomic Charge', 400.
[20] H. G. J. Moseley, in *Nature* **92** (1914), 554. 'My work was undertaken for the express purpose of testing [van den] Broek's hypothesis ... [and] certainly confirms the hypothesis'.
[21] 'Soddy's prediction concerning the atomic weights of leads from uranium and thorium minerals had been triumphantly vindicated by some of his most severe critics'. F. W. Aston, 'The Story of Isotopes', in *British Association Report* (1935), Presidential Address to Section A, p. 26. The concurrent investigations comparing uranium lead with ordinary lead could neither confirm nor deny the possibility of thorium lead.
[22] Brewer, 'Chemistry at Oxford', 185.
[23] Soddy, 'The Internal Energy of Elements', in *Journal of the Proceedings of the Institution of Electrical Engineers, Glasgow* **37** (1906), 7. An earlier statement on the latent internal energy of the atom is Soddy, 'The Disintegration Theory of Radioactivity', in *Times Literary Supplement* (26 June 1903), 201.
[24] Soddy, 'Advances in the Study of Radio-active Bodies', two lectures to the Royal Institution on 15 May and 18 May 1915, as recorded in *The Royal Institution Friday Evening Lectures 1907–1918* (privately bound at the Royal Institution, London, n.d.). The original MS is in the Bodleian Library, Soddy-Howorth Collection, 58. The quotation is from this MS, page **II**, 9. The lectures are apparently unpublished but are reviewed in *Engineering* **99** (1915), 604.
[25] Shortly after Soddy's retirement, Joseph Needham pointed out the importance of such efforts, 'Social Relations of Science', in *Nature* **141** (1938), 734.
[26] Soddy, *The Frustration of Science*, Foreword.

[27] Soddy, Typescript-A, 1953, concluding statement, Bodleian Library, Soddy-Howorth Collection 4.

## Bibliography

A nearly complete list of Soddy's main scientific papers, books, lectures, and other contributions is given by Alexander Fleck, 'Frederick Soddy', in *Biographical Memoirs of Fellows of the Royal Society* **3** (1957), 203–216. For comparisons and additions, including his contributions on economics and on science and society, see Muriel Howorth, *Pioneer Research on the Atom* (London, 1958), 281–286. This unusual account is subtitled *Rutherford and Soddy in a Glorious Chapter of Science*, and further subtitled *The Life Story of Frederick Soddy*. In spite of the author's uncritical attempt to glorify Soddy, this remarkable reference source is the fruit of great effort to preserve the existing documents of Soddy. Soddy gave all his papers to Muriel Howorth of Eastbourne, and his will contained the provision: 'I give to Muriel Howorth also the copyright of all my published works', cf. *Pioneer Research*, p. 286. The Soddy-Howorth Collection was deposited in the Bodleian Library and a partial reference key thereto is appended to *Pioneer Research*, pp. 333–339. In 1974 J. Alton of the Contemporary Scientific Archives Centre, Oxford, deposited in the Bodleian a 29-page systematic catalogue of the Soddy Collection incorporating the Howorth portion. This collection must be directly consulted for precision in both quotations and other references. Richard Lucas, *Bibliographie der radioaktiven Stoffe* (Leipzig, 1908), 72–73, provides a useful list of Soddy's early works. Consultation of the British Museum General Catalogue of Printed Books, 1964, amplifies the list of works of Soddy. In addition to the scientific contributions collectively listed in Fleck and Howorth, the following should be noted: 'The First Quarter-Century of Radioactivity', in *Isotopy* (Westminster, 1954), 1–25. See the obituaries of 'Rutherford', in *Nature* (30 October, 1937); 'Ramsay', *ibid.* (10 August 1916); and of H. Becquerel, 'The Founder of Radioactivity', in *Ion: A Journal of Electronics, Atomistics, Ionology, Radioactivity and Raumchemistry* **1** (1908), 2–4. Soddy was joint editor of this short-lived serial, *Ion*. In this same issue, Soddy completed his series of investigations concerning whether the alpha particle was charged before, during, or after expulsion. Soddy's abstracts of the papers by Russell, Fajans, and Soddy concerning the displacement law are also of interest: see *Abstracts of Chemical Papers Journal of the Chemical Society London*, pt. 2 (1913), 274–278. Soddy's classic call for scientific responsibility appears as the foreword to *The Frustration of Science* (London, 1935).

Soddy's nine joint papers with Rutherford (1902–1903) are reproduced in *Collected Papers of Lord Rutherford of Nelson*, Sir James Chadwick (ed.), I (London, 1962). Soddy contributed a series of original reports on 'Radioactivity' for the *Annual Reports on the Progress of Chemistry* (London, 1904–1920). These articles contain much otherwise unpublished work on isotopes, as well as a running account of the history of radioactivity. These articles have been published in facsimile and edited with commentary by T. J. Trenn, in *Radioactivity and Atomic Theory* (London, 1975). The diagram 'Radio-Elements and Periodic Law' first appeared as a supplement to Soddy's paper 'The Radio-elements and the Periodic Law', in *Chemical News* **107** (28 Feb. 1913), 97–99. Essentially the same diagram appeared in *Jahrbuch der Radioaktivität und Elektronik* **10**, No. 2

(1913), 193. The actinium series was separated and minor additions were included in the version drafted July 1913 for the *British Association Report* (1913), 446, and here reproduced; it also appeared in the *Annual Report* **10** (1913), 264.

Soddy's most important books are *Radio-Activity: an Elementary Treatise From the Standpoint of the Disintegration Theory* (London–Leipzig, 1904), based upon a series of lectures at the University of London from Oct. 1903 to Feb. 1904, carried concurrently in *The Electrician*, **52** (1903), 7 et seq.; *The Interpretation of Radium* (London, 1909; 4th ed., 1920), translated into several languages. In the series edited by Alexander Findlay, *Monographs on Inorganic and Physical Chemistry*, Soddy contributed *The Chemistry of the Radio-Elements*, pt. I (London, 1911; Leipzig, 1912); pt. II (1914), containing 'Radioelements and the Periodic Law'; and pt. I, 2nd ed. (1915). See also *The Interpretation of the Atom* (London, 1932).

Soddy's most important lectures were The Wilde Lecture VIII, 'The Evolution of Matter as Revealed by the Radioactive Elements,' 16 March 1904, in *Memoirs and Proceedings of the Manchester Literary and Philosophical Society* **48** (1904: Leipzig, 1904); The Nobel Lecture, 12 Dec. 1922, 'The Origin of the Conception of Isotopes', in *Les Prix Nobel en 1921–1922* (Stockholm, 1923).

Information concerning the life and work of Soddy can be obtained from *Pioneer Research*. Howorth also edited the *Memoirs* of Soddy, as *Atomic Transmutation, Memoirs of Professor Frederick Soddy*, vol. I (London, 1953), subtitled *The Greatest Discovery Ever Made*. Volume one deals with the period until 1904. There were no further volumes. There are numerous sketches of Soddy's life and work. Alexander Fleck, in *Nature* **178** (1956), 893, is an interesting personal account. Fleck also contributed the note for the *Dictionary of National Biography* (1951–1960), 904. Alexander S. Russell, 'F. Soddy, Interpreter of Atomic Structure', in *Science* **124** (1956), provides insights into Soddy the man. Russell published further on Soddy, in *Chemistry and Industry*, No. 47 (1956), 1420–1421, and in Eduard Farber (ed.), *Great Chemists* (New York, 1961), 1463–1468. Perhaps the best account is F. Paneth, 'A Tribute to Frederick Soddy', in *Nature* **180** (1957), 1085–1087; repr. in the Paneth Collection, H. Dingle (ed.), *Chemistry and Beyond* (London, 1964), 85–89. A more recent sympathetic account is that of A. Kent, 'Frederick Soddy,' in *Proceedings of the Chemical Society* (November 1963), 327–330. Besides his brief editorial 'Frederick Soddy and the Concept of Isotopes,' in *Endeavour* **23** (1964), 54, T. I. Williams wrote the article on Soddy for his *Biographical Dictionary of Scientists* (London, 1969). The account of I. Asimov, *Biographical Encyclopedia of Science and Technology* (New York, 1964), no. 398, is subject to the limitations imposed by this effort. An extremely concise and accurate summary is included in W. A. Tilden and S. Glasstone, *Chemical Discovery and Invention in the Twentieth Century* (London, 1936), 140. There is a supplementary account in Eduard Farber, *Nobel Price Winners in Chemistry 1901–1961* (London, 1963), 81–85. It is of interest to compare the biographical account in *Nobel Lectures in Chemistry* (Amsterdam, 1966), 400–401, with the original in *Les Prix Nobel en 1921–1922* (Stockholm, 1923), 128–129. See also the account in H. H. Stephenson, *Who's Who in Science* (London, 1914), 535, and *Journal of Chemical Education* **8** (1931), 1245–1246.

Relevant sketches of Soddy's work are to be found in F. W. Aston, *Isotopes*, 2nd ed., 1924. His work on lead isotopes, pp. 17–19, is particularly valuable. See A. Kent and J. A. Cranston, 'The Soddy Box,' in *Chemistry and Industry* (1960), 1206, 1411, which describes Soddy's original 1910 preparation, a deliberate mixture of radium and mesothorium, which led him to the concept of the isotope. In Gleditsch, 'Contribution to the Study of Isotopes,' *Norske Videnskaps-Akademi* I. *Mat.-Natur. Klasse* No. 3 (Oslo, 1925), E. Gleditsch notes, p. 7, that 'The theory of isotopes put forward . . . by Soddy in the years 1911–1914 has proved to be fully in accord with our present views on atomic structure'. See also Fleck, 'Early Work on the Radioactive Elements,' in *Proceedings of the Chemical Society* (1963), 330. In this same issue, J. A. Cranston contributed 'The Group Displacement Law,' pp. 330–331, and an even more detailed documentation in the following issue (1964), 104–107. Soddy's work with Rutherford is considered by A. S. Eve, *Rutherford* (Cambridge, 1939); N. Feather, *Lord Rutherford* (London, 1940); A. Romer, *The Restless Atom* (New York, 1960); Howorth, *Pioneer Research*; and T. J. Trenn, 'Rutherford and Soddy: From a Search for Radioactive Constituents to the Disintegration Theory of Radioactivity,' in *Rete* **1** (1971), 51–70. M. W. Travers, *A Life of Sir William Ramsay* (London, 1956), Chapter 14, pp. 210–221, deals with his work with Ramsay. At the request of Travers. Soddy contributed a portion of this account. The original transcript is in Soddy-Howorth 4.

In addition to F. M. Brewer, 'The Place of Chemistry at Oxford,' in *Proceedings of the Chemical Society* (July 1957), 185. Soddy's work at Oxford is considered by Sir Harold Hartley, 'The Old Chemical Department,' in *Journal of the Royal Institute of Chemistry* (1955), 126. J. A. Cranston's 'The Discovery of Isotopes by Soddy and his School in Glasgow,' in *Isotopy* (1954), 26–36, and 'Concept of Isotope,' in *Journal of the Royal Institute of Chemistry* **18** (1964), 38, provide important historical and scientific distinctions in the use of the term 'isotope.'

A. Romer (ed.), *The Discovery of Radioactivity and Transmutation*, Classics of Science, II (New York, 1964), provides not only some of the papers of Soddy in collaboration both with Rutherford and with Ramsay but also valuable comments on this pre-1904 work. Soddy's hypothesis concerning an isotope of lead as the final product of the thorium series is dealt with in S. I. Levy, *The Rare Earths* (London, 1915), 107–108.

For a partial account of Soddy's work on isotopes, emphasizing the contributions of Fajans and Richards, see O. U. Anders, 'The Place of Isotopes in the Periodic Table: the 50th Anniversary of the Fajans-Soddy Displacement Laws,' in *Journal of Chemical Education* **41** (1964), 522–525. Additional information about Soddy as others saw him is in L. Badash (ed.), *Rutherford and Boltwood: Letters on Radioactivity* (New Haven, 1969), which exposes Soddy's research on the parent of radium. Soddy as a public figure and social rebel, who ushered in the atomic age, is epitomized in C. Beaton and K. Tynan, *Persona Grata* (London, 1953), 87. Besides the Soddy-Howorth Collection, extensive correspondence exists also at the Cambridge Univ. Library, Add. MSS 7653/S. There is also correspondence with W. H. Bragg at the Royal Institution, with J. Larmor courtesy the Royal Society, and with O. Lodge at University College, London. The Soddy Memorial at Glasgow was reported in 'Unveiling of the Soddy Memorial,' in *Chemistry and Industry* (8 Nov. 1958), 1462–1464. There is one collection of Soddy's apparatus and equipment at the Chemistry Department of the University of Glasgow and another at the Inorganic Chemistry Laboratory of the University of Oxford. A history of the Rutherford – Soddy collaboration is given in Thaddeus J. Trenn, *The Self-Splitting Atom* (London, 1977).

# PART I

# SELECTED PAPERS OF SODDY'S

FREDERICK SODDY

# TRANSMUTATION
# THE VITAL PROBLEM OF THE FUTURE*

The advances of the last fifteen years consequent upon the discovery of the property of radioactivity by M. Becquerel in 1896, and of the separation from uranium minerals of the new element radium by M. and Mme. Curie have brought many changes of outlook upon science. Isolated questions of the most extraordinary diversity have been profitably reconsidered. It is only necessary to cite at random a few of the special problems in individual sciences, to which the new knowledge has been applied, to illustrate the fruitfulness of the new subject in realms unconnected and apparently far remote from its own. Some of these, for example, are the origin of the pleochroic halos in certain minerals, the medicinal efficacy of certain spas, the occurrence of explosive gases in certain mines other than those from which coal is gained, the colours of gems, the high temperature gradient of the Simplon tunnel and the electrification of the atmosphere in the neighbourhood of mountain tops. The explanation of all the various manifestations of energy associated with radioactivity has been found in the discovery of naturally occurring processes of transmutation progressing continuously and invariably within the radio-elements. Of these new fundamental sub-atomic processes, all the varied phenomena of radioactivity, – the spontaneous and continuous emission of light and heat, the expulsion of $\alpha$-, $\beta$-, and $\gamma$-rays and of radioactive 'emanations', the generation of electricity and the energetic chemical reactions, – are merely incidents. Moreover they can scarcely be considered necessary or essential incidents, for processes of the same type, if but slightly less energetic, might well occur in other elements without there being any evidence that would come within the known methods of detection. Radioactivity has been the means whereby it has been possible for the first time experimentally to investigate the process of transmutation as it occurs in Nature. But this process of transmutation, now it is a proved experimental reality, carries with it consequences far beyond the domain of radioactivity or even of physical science itself. It contains suggestions and applications of the greatest general interest connected with the past history and future destiny of this world and of the race it bears. It offers novel information con-

*George B. Kauffman (ed.), Frederick Soddy (1877–1956)*, 3–17.

cerning of the manner in which Nature stores and draws for its available supplies upon that fundamental commodity, energy. A constant and never-ending fresh supply of energy is the major physical condition which limits the preservation of life in any form. Moreover an increasing and ever-more increasing supply of this commodity must be available as civilisation expands and disentangles the wheels of human destiny from the sport and caprice of external circumstances.

A report of a find of gold in any part of the world is sufficient to start a pilgrimage thence of all and sundry seeking the otherwise rather overrated precious metal of which the coins or counters of the world's commerce are, by convention, composed. It hardly needs an economist to appreciate the fact that the commodity so mined is not wealth, but merely its conventional symbol, adopted for convenience as a medium of exchange. On the other hand an economist might not at once be prepared to accept the doctrine, to which physical science of the last half-century plainly leads, that so far as human affairs go, wealth and available energy are synonyms, and that the poverty or affluence of this planet are primarily measured only by the dearth or abundance of the supply of energy available for its life and work. In use this energy is not destroyed but is transformed into useless or waste forms, such as heat of uniform temperature, which hardly need be further considered in the present connection. Of course there are degrees also in the intrinsic value of available energy. The chemical energy of coal and other forms of fuel may be considered the standard form of available energy. The value of the energy of waterfalls varies much more than that of coal according to the local conditions, such as the continuity of the supply during periods of drought. The enormous supplies of available energy associated with the tidal movements of the ocean, derived primarily from the energy of the earth's rotation, are practically valueless because of their intermittent availability and the enormous volumes of water, the movement of which must be controlled if the energy is to be utilised.

The radioactive study of transmutation has revealed in matter an enormous and previously unsuspected store of energy compared with which all previously known sources of energy shrink into the merest insignificance. At present, however, this store is entirely valueless, or, at least, its value is entirely prospective and dependent upon further discovery. Unlike tidal energy its valuelessness does not depend upon the enormous quantity of matter associated with relatively small amounts of energy. Quite the contrary is the case. The energy in radium is, weight for weight of matter, of the order of a million times the energy of coal or other fuel. Neither is its valuelessness due to its inconvertibility into useful forms in the same sense as in the case of the waste energy of heat of uniform temperature. Even with the small amounts of radium available more than one 'perpetual motion machine' has been constructed converting a small fraction of the energy evolved into mechanical movement. The difficulty is entirely connected with the slow rate at which these newly discovered stores of energy in matter are naturally evolved. We know, for example, that radium will continue to give out its energy at hardly diminished rate for hundreds and even thousands of years. In the sum total the energy evolved is enormous, being, as already stated, of the order of a million times that evolved in the combustion of a similar mass of coal. But over a few hours or days the energy, though amazing eough in many of its effects, is in actual quantity insignificant for useful purposes. To make it and similar stores of energy in common matter of use we must be able to hasten the natural process at will. This is the unaccomplished step, the passage from the passive study of a natural process of atomic transmutation, which the radio-elements are undergoing, to the active achievement of transmutation on our own account. The ancient problem of transmutation has thus in the last decade taken on an entirely new aspect. There is abundant reason for believing that these new stores of internal atomic energy, which in the case of the radio-elements are being slowly and spontaneously evolved, are the fundamental natural stores of energy, and that all or almost all the available energy of nature is derived ultimately from these storehouses during the long epochs of cosmical evolution.

It may appear strange at first sight that a science, so novel and youthful as radioactivity is, should be in a position to put forward generalisations of such definitness and finality, which in many ways shed an entirely new light on questions of fundamental interest. There are probably two reasons for this. It has been the good fortune of the science, in the first place, to have attracted investigators, at a very early stage, of unquestioned genius and extraordinary energy, so that from the very first it has been directed by leaders at once capable of directing it into fruitful channels and of exposing incontrovertibly such false steps as have been made, before they have had time to become established. In the second place there is overshadowing the whole subject an atmosphere of the absolute and the inevitable, rare even in any other branch of physical science. An experimental fact is true or not true in a most unqualified sense. In the field of chemical changes, for example, slight alterations, often of an unknown character, of the conditions of the experiment and of the environment in which a phenomenon occurs produce the most profound differences in the results obtained. In radioactivity it is never so. The phenomena if they occur at all, occur always at the same rate and with the same result. It is, at present, entirely beyond the power of the investigator to influence in any real way the progress of the phenomenon he is investigating. Naturally most exhaustive experimental trials have been made to discover within what limits, if any, this detachment of the processes of radioactivity from the influence of its external environment holds good. As a result it is established that within the limits of conditions at the disposal of science to-day, under conditions of temperature, pressure and so on available in a modern laboratory, it is beyond the power of the investigator to influence in any real way the progress of the phenomena he is investigating. In consequence it is rightly claimed that when an experimental fact has been established, it must be accepted as of general unqualified application, at least until the contrary has been proved. It must necessarily always involve an assumption to apply knowledge gained as the result of experiment under laboratory conditions to the transcendental conditions prevailing elsewhere in the universe. But if such an assumption can ever be made it can be made most legitimately in the case of the phenomena of radioactivity. For this reason the application of the new discoveries to the wide questions of the source and flow of energy throughout the universe, is not entirely hypothetical.

A few words of introduction as to what is accepted as proved among all the workers in radioactivity, practically without exception, may usefully precede the consideration of the problem of artificial transmutation. A radio-element is now recognised as an element, the atoms of which are in progress of disintegration. Each ray emitted – and it is possible to detect the expulsion of an individual $\alpha$-ray by its phosphorescent effect, as in the spinthariscope, or by its electrical effect in ionising gases – corresponds in general to the sudden disintegration of a single atom of the radio-element into two parts. The lighter part is expelled with a velocity hitherto unknown for matter as a radiant particle, or 'ray', while the heavier part or residue of the original atom, recoils with a similar momentum, and in consequence far smaller velocity. This recoil can easily be detected by suitable experiments. In the majority of cases the 'ray' expelled consists of an atom of the element helium ($\alpha$-rays), in other cases ($\beta$-rays) of an electron or atom of negative electricity disembodied, so to speak, and travelling unencumbered by matter. A third type of ray, known as the $\gamma$-ray, occurs as the accompaniment of the expulsion of the $\beta$-ray, but its physical nature is not yet clearly decided. As a general rule the heavier residue of the disintegrated atom, disintegrates again in a definite manner after a definite average period expelling a new ray and producing a second residue, which proceeds exactly in the same way to pass through a third and fourth disintegration. In addition to the helium and the expelled electrons there exists in consequence a series of unstable intermediate forms of matter, distinguished by remaining in existence only a limited time, which on the average is for each form a definite period, known as 'the average life', and which can in most cases be exactly determined. These periods of average life constitute a series of remarkable fundamental natural time-constants, varying from fractions of a second, on the one hand, to geological epochs of time, on the other. The considerations already referred to, according to which it is impossible to alter the natural rate at which a radio-element disintegrates, make these new time-constants more absolute and fundamental than any other standards of time. As is well known, our standards of time, which are based on astronomical phenomena, are not absolute but suffer secular variation. A system of time-units based upon the periods of average life of the various products of radioactive changes would, there is every reason to believe, be free from this defect and could easily be reproduced *ab initio* in any world where any of the radio-elements existed. A short description of how these fundamental constants are determined may therefore be of interest. The determination is based upon the fact that the changes of all radio-elements follow the same general law. A definite fraction of the total amount of radio-element undergoes change in each unit of time, and this fraction is denoted by the symbol $\lambda$, and termed the radioactive constant. In the language of the differential calculus

$$\frac{dQ}{dt} = -\lambda Q$$

where $Q$ represents the quantity of the radio-element in existence. The first consequence of this is that the quantity, provided some definite amount $Q_0$ is taken to start with, diminishes exponentially with the time, or, as it is sometimes expressed, as the time increases in arithmetical progression the quantity diminishes in geometrical progression. If after the first unit of time $\frac{1}{2}Q_0$ has changed and $\frac{1}{2}Q_0$ remains unchanged, after the second unit of time $\frac{1}{4}Q_0$ remains unchanged, after the third $\frac{1}{8}Q_0$ and so on. In general if after the lapse of unit time the fraction $q$ remains unchanged, then, after the lapse of $t$ units of time, the quantity remaining unchanged is $(q)^t$. The constant $\lambda$ of the former equation is the fraction of the whole changing in the unit of time, provided the latter is chosen so small that the total quantity does not change perceptibly during that time, or what amounts to the same thing, provided the quantity were maintained at its initial value during the time-unit from an external source.

The next mathematical consequence is that if at any instant, reckoned as the starting point, a certain number of atoms of the radio-element are considered, the average period of time each atom remains in existence before it changes is definite and is equal to $1/\lambda$. The actual periods of the individual atoms assume all values from zero to an indefinitely long period, but the average period is definite. So this average period is known when the constant $\lambda$ is known, and the latter is easily determined for quickly changing substances by observing the decay of the radioactivity with lapse of time. For slowly changing substances, such as radium, the fraction changing and consequently the decay of the activity, in any available period of time is too small to be determined, and another method is used based upon a third mathematical consequence of the law of radioactive change when applied to a series of successive similar changes. If, as must of necessity be the usual case in nature, the first change is the slowest of the series, the relative quantities of all the members of the series of changes assume a definite ratio to the quantity of the parent element. The relative quantity of each member is proportional to its period of average life. This is known as the state of radioactive equilibrium. One of the radioactive disintegration series starts with uranium and the most important successive products of uranium, in order of sequence, are ionium, radium, radium emanation, radio-lead, polonium. The relative quantities of all these substances in a mineral containing uranium, provided that all are retained by the mineral, are proportional to the periods of average life of the substances. By taking one of the shorter lived members, such as the emanation or polonium, the period of which is known from direct observation, and determining the ratio of its quantity in a mineral to that of the other longer lived and therefore more abundant members, it is possible to determine the periods of the latter, although these may be extremely great. The method naturally must be used with caution and is subject to certain limitations of accuracy which need not be here detailed, but on the whole there is no reason to distrust the general correctness of its results. In this way, for example, and by independent methods, it is found that the period of average life of radium is about 2500 years and that of uranium about 3 000 000 times greater, the quantity of radium in minerals being nearly constant at about 1:3 000 000 in terms of the quantity of uranium. The period of average life of uranium is therefore about 7500 million years and since this is a long period compared even with the longest demanded by the requirements of geology for the age of the earth, there is no immediate necessity of enquiring as to the origin of the parent uranium. Of this we have no present evidence. It must suffice that there is no inconsistency in assuming that the scheme of changes disclosed by radioactivity may have been proceeding uniformly at the present rate in the earth for periods far longer than the period over which we have evidence that the earth itself has been in existence in its present form. At the other extreme are certain very short-lived products, such as the thorium emanation and the actinium emanation, the average periods of life of which are to be estimated in seconds, and which therefore, apart from their longer lived parents, can have only a very brief existence. But since they are being continually reproduced in the changes of these parents, they can be studied quite well in passing, so to speak. As a matter of fact they are almost as easily and completely investigated as the more stable types. It is only when the period of life is a short fraction of a second, as in some cases now being examined, that the investigation of these ephemeral elements becomes difficult.

The same general considerations apply equally whether the unstable element is, like uranium, changing with infinite slowness, or whether its existence as a separate element is brief beyond the ordinary methods of measurement. While they last these new unstable elements are in no way distinguishable from ordinary elements. They possess definite chemical and physical properties, which are often somewhat difficult to determine on account of their necessarily almost infinitesimal quantity. Notwithstanding this, however, a surprising amount of accurate information has been accumulated with reference to the ordinary material properties, the volatility, solubility, electrochemical position and general chemical characteristics of the twenty or more relatively short-lived products of radioactive changes which are now known. In all about

30 radio-elements are known, but 22 have periods less than one year.

Radium is recognised as one product of the disintegration of uranium, and polonium a later product in the same series produced from radium. The periods of average lives of these two substances are 2500 years and 200 days respectively. For the few thousands of years, which, on the average, its atoms exist, radium behaves exactly like a common element, being in fact the heaviest member of the well known alkaline-earth family. It has a definite spectrum and atomic weight, and a chemical nature very closely allied to, but distinct from, barium. While it lasts, it is an inactive element like barium. The peculiar radioactive properties of radium only occur at the moment its individual atoms cease to exist and disintegrate into new atoms. But since in any perceptible mass of radium there must be myriads of single atoms, and a small definite fraction of these breaks up every instant, the radioactivity *appears* as the *continuous* property of the whole mass.

The most difficult part in the elucidation of these disintegration series is to establish the identity of the final product. The last known change of the uranium series is that of polonium and the nature of the final product other than the helium ($\alpha$-particles), which polonium has been proved to produce, has not yet been directly proved. But there is every reason for believing that it will turn out to be the common element lead, when sufficient of it has been accumulated for chemical and spectroscopic identification.

The intermediate products, even the existence of this remarkable series of changes itself, would have remained unknown and unknowable were it not that in these newly recognised atomic changes enormous quantities of energy are evolved. The energy given out by any mass of radium per day, during which time little more than one millionth part of the whole disintegrates, is comparable in amount with what would be given out from a similar mass of matter in the most energetic chemical changes known. The energy of sub-atomic changes is thus of the order of a million times greater than the energy of ordinary chemical or molecular changes. The energy evolved from the disintegration of a single atom is detectable by radioactive methods, whereas a million million atoms of any non-radioactive element is a quite undetectable quantity even with the spectroscope. For this reason the identity of the final non-radioactive product of the whole sequence of changes remains still uncertain.

Radioactive change is a spontaneous and natural process of transmutation, taking place under all circumstances in the case of the heaviest known element uranium, and of the next heaviest known, thorium. The fact that it has been found quite impossible, even by the most powerful agencies known, to alter artificially the rate at which a radioactive substance is changing, either to retard or to accelerate it, is obviously the corollary to the well known impossibility of artificially transmuting one element into another. Since, even where one element is naturally changing into another, all attempts to influence in the slightest degree the rate of the process have failed, it is not to be expected that it would be possible to institute a transmutation at will. Least of all is it conceivable that the early alchemists with their relatively far less powerful appliances were ever successful in their quest. On the other hand, if it were possible, by alteration of the external conditions, to alter the rate at which a radio-element undergoes transformation, or, in other words, if it were possible to alter the degree of radioactivity of a radio-element at will, that would be tantamount to artificial transmutation and there is little doubt that the transmutation of other elements, not radioactive, would soon follow. The modern scientific assault on this problem of transmutation lies therefore through the study of radioactive processes. It may, without fear of question, be asserted that this problem, to-day apparently so remote and academical, and so far from solution, is the physical problem towards which the civilisation of man is furiously hurrying him, and upon which, if still unsolved, his newly assumed control over Nature will be decisively tested to destruction.

The world has advanced far in thought since the possibility of being able to convert a common metal into gold was regarded as the highest aspiration of knowledge. The special value of gold is, of course, merely conventional and could and would be changed by a stroke of the legislator's pen. But, as Faraday remarked in his day, transmutation is the final goal towards which chemistry should aspire. The power to decompose and build up the known elements and to construct new ones at will as is now done for compounds, would elevate chemistry to an infinitely loftier plane than the rather secondary and subordinate position among the physical sciences it occupies at the moment.

Even so, it is not at all on the material or chemical side that the greatest consequences of the solution of the problem of transmutation would accrue. Energy, not gold, will be the quest of the modern scientific alchemist. It is a well recognised principle of physical chemistry that the absolute amount of energy contained by any mass of matter is unknowable. All that can be determined is the difference in the amount before and after the matter has gone through any change. The differences in the energy contents of an element and its compounds are well known and form the basis of thermo-chemistry. The difference in the energy contents of gold and silver is unknown, because gold cannot be converted into silver or *vice versa*. But by the study of the change, for example, of uranium into radium, or of radium in turn, into lead, since the assumption made appears well justified, we have for the first time the means of determining the difference between the absolute energy contents of pairs of elements, and enormous it proves to be. It is of the order of a million times greater than the quantities met with in thermochemistry. In the disintegration of a gram of uranium, a process which is half completed in a period something like five thousand million years, the heat evolved would be of the same order as that produced in the combustion of a ton of coal. In consequence, to reverse the process and to build up artificially a gram of uranium from lead, there must be added, in addition to helium (eight atoms of which are evolved from each atom of uranium during disintegration) the energy contained in at least a ton of coal. So far, knowledge is confined to the one case of natural transmutation, that of the heavier elements into ones of lesser atomic weight with evolution of energy. It seems natural to assume that the elements of greater atomic weight possess, generally, greater intrinsic internal energy than the lighter ones. If this be so, the transmutation of silver into gold would be, in all probability, a ruinous process. It is probable that, if ever transmutation becomes practically possible, the material products of the process will be of little more account than the products of combustion of the furnaces of the industrial world to-day. With energy at its present value, no known elementary substance, probably, is so valuable that it could be built up with profit from elements of less atomic weight, owing to the enormous expenditure of energy required, or that could not with profit be disintegrated into lighter forms, were this possible, for the sake of the energy that would be obtained. The modern scientific alchemist would probably be more eager to succeed first in turning gold into silver than the opposite.

If the problem of transmutation is ever solved, it is to be expected that the synthesis of entirely new forms of matter will also be possible. Engineering and allied sciences in the great development of the past few decades, have requisitioned many of the elements hitherto thought to be rare and of no technical application. The cases of the use of thorium in the manufacture of the incandescent gas-mantle, and of all kinds of rare elements in the composition of alloys for special purposes, are illustrations which could be multiplied indefinitely. In this respect the progress of science is at present absolutely limited by the elements available. In the present state of the science of aviation it is difficult to believe that an entirely satisfactory solution will be arrived at, until a constructional material as light as wood and as strong as steel is available, and the choice of light elementary materials at the disposal of the chemist is so limited that probably no great advances can be expected in this direction. But if new elements could be prepared to a definite specification by artificial synthetical processes in the same way as chemical compounds are now turned out of almost any structure required, within certain limits, this fundamental limitation of further progress would be overcome.

Such would however be but a side issue of the main result, which would place at the disposal of science a new and practically inexhaustible fund of energy incomparably more abundant than any at present available. Even the partial solution of the problem, as far as the acceleration at will of the natural rate of disintegration of one of the two parent radio-elements, uranium or thorium, would bring about an enormous revolution, for both of these elements are sufficiently common and widely distributed to supply all the energy the world is likely to be able to use for a very long period. For every million tons of coal burnt to-day a single ton of either of these elements, under the new circumstances, would suffice. The world's annual consumption of thorium for the gas-mantle industry amounts to several hundreds of tons.

Even among scientific men the general modern method of treating energy as a fundamental commodity is often obscured by the survival of less direct and more special habits of thought. Sir William Crookes in a remarkable address to the British Association some years ago considered the great question of the world's future supply of wheat and drew the somewhat disturbing conclusion that a limit must soon be reached to possible further expansion in this direction, and that then a wheat famine would be inevitable. It happens that in this particular case we have an easily understood and specially glaring instance of what is really a universal question. For wheat is only energy associated with some of the commonest and most abundant chemical elements, by means of the ancient, still not perfectly understood, and doubtless highly laborious and inefficient methods of agriculture. The source of the energy in this case is the sun. It is supplemented in modern agriculture by manuring the soil, the most valuable constituents of the manure being nitrogen in the combined form, that is to say, nitrogen either in combination with hydrogen as ammonia and its compounds, or with oxygen as nitrates, etc. In this complex series of changes it is apt to be overlooked that the combined nitrogen is essentially nitrogen plus energy, not nitrogen plus hydrogen, or nitrogen plus oxygen, as it is usually regarded, and that the abundant and almost inexhaustible supply of atmospheric nitrogen as such is practically unavailable for agricultural purposes because it lacks energy. Atmospheric nitrogen is 'self-combined nitrogen', for the individual atoms of elementary nitrogen are combined in pairs to form nitrogen molecules and in this combination there is evolved an enormous quantity of energy. To make atmospheric nitrogen available this energy must be returned. In Nature there is a very slow but continuous process going on, transforming the self-combined nitrogen of the air into 'combined nitrogen' i.e. ammonia and nitrates. It has been supposed that the sun was the source of the energy required in this transformation, which was effected by means of lightning flashes and silent electrical discharges. It seems much more probable however that the energy is really derived from the radioactive materials in the atmosphere, the rays of which cause the combination of atmospheric nitrogen and oxygen with ease. This is however a big question and cannot be dealt with here. This process, which sufficed for primitive agriculture, is insufficient to-day, and in consequence, combined nitrogen – ammonium salts from gas-works and the Chilean deposits of sodium nitrate – are supplied to the soil. Now the Chilean deposits happen to be the only ones of their kind and are certainly not inexhaustible. When they are used up one of the vital assets of agriculture will have disappeared, and thus the 'wheat problem', a phase on a small scale of the still unrecognised 'energy problem' that awaits the future, is the first to have attracted attention.

The sequel is perhaps even more instructive. Great commercial corporations, content to work for years or even decades, spending great sums in the hope of a future harvest, quickly grasped the inwardness of the 'wheat problem'. There are more direct sources of energy than the sun's light and heat. Such are, for example, coal and waterpower. The conversion of atmospheric nitrogen into combined nitrogen was effected by Cavendish in the 18th century by means of an electrical machine, the energy of which was supplied by himself and his assistant, taking turns at the wheel. In the modern industrial development, known as 'the utilisation of atmospheric nitrogen', the wheel, in this case the modern dynamo, is turned by the water power of Norwegian hill sides or by steam turbines driven by coal. The problem is one of engineering rather than of pure science. There is little doubt that by the time the Chilean deposits are worked out the great scientific genius of the 20th century will be credited with having created an even more satisfactory substitute.

Modern science, and its synonym modern civilisation, however create nothing. Their raw material is energy, and this though immaterial in the literal sense, is vital in the most fundamental sense it is possible to use the word. Life came and slowly developed on this planet at a rate allowed by the day-to-day supply of natural energy, subject to its fluctuations, imperilled by every drought or other abnormal condition affecting the intricate ramifications of the supply. Then science, still popularly misunderstood as a creator, developed, and at each stage of its growth it has drawn on some otherwise valueless store of energy for its raw material. It is ransacking the globe of stores which have required geological epochs of the past to lay down, and which itself it is powerless to replace.

Coal must follow ultimately the threatened fate of the Chilean nitrate beds and considering the rapidly increasing rate at which the available resources of energy are being used up, the end may come sooner than even the most pessimistic anticipate. The popular idea that when the coal supply is exhausted 'the great scientific genius of the *n*-th century will have provided an even more satisfactory substitute' is pernicious. Energy cannot be created and the utmost accomplishment of science, so far as can be seen, is to use to the best advantage what Nature has provided.

It is possible, of course, to quibble over the exact term of years the visible energy resources of the globe may be expected to last, but of their rapid ultimate exhaustion it does not seem possible to doubt. It is true that as the surface supplies give out fresh supplies may be found at greater depths, but there is an obvious limit, when the energy that must be expended in deep mining becomes comparable with the energy obtained.

As regards energy, and therefore as regards every other commodity, the modern world is undoubtedly living far beyond its income. It has recently come into a legacy from the remote past and it is living on the capital. It cannot now be very long before it wakes up to the appreciation of this fact.

The main discovery of radioactivity, therefore, that elements are capable of undergoing transmutation, the one into the other, with the liberation of hitherto unsuspected quantities of energy, has come at a most opportune moment in the sequence of events. Under existing conditions civilisation must come to an end the sooner the more rapid its progress is and the greater the heights to which it attains, for present day civilisation might be described as due to the supplementing of available supplies of natural energy by means of the accumulated stores of byegone ages. These stores may last out a century or two longer, but it is obvious that the more glorious the zenith attained the more swift and sure will be the decline. The only possible way of escape known to science is by the solution of the problem of transmutation. The thoughtless may conclude lightly that science which has accomplished so many marvels will surely not fail when this ultimate test comes. But the scientific man and particularly the chemist is under no delusions as to the nature of the problem and how completely it transcends in order of magnitude any that have been previously solved. Physical science so far has dealt with the mere external properties of atoms, each of which has been a sealed book to all methods of enquiry. At one time it was thought that it would be possible, by means of the spectroscope and by the behavior of gaseous matter in highly rarefied state under the electrical discharge, to obtain information as to the internal constitution and properties of the atoms. But such a belief is rapidly losing ground and it is recognised that while these methods may penetrate slightly further than others they deal with at most the outer fringe. The kernel of the atom still remains impenetrable, and it is only in radioactivity that it comes into evidence. The most recent development of this subject may possibly prove to be the beginning. The $\alpha$-particles are atoms of helium expelled with the velocity of about one-fifteenth that of light, and this enormous velocity is sufficient to enable the $\alpha$-particle to pass rectilinearly and completely *through* many atoms of matter in its path before its initial energy is spent. These $\alpha$-particles thus have access to the innermost portions of the atoms of all matter, and it remains to be seen whether, no emergence therefrom, they may not provide some useful clue as to the nature of the unknown territory they have traversed. However that may be, it may be said, without much fear of denial, that the whole range of physical science outside of radioactivity is admittedly of little or no help in the problem of artificial transmutation. Each atom remains as it were an unknown inland ocean whilst the existing science of matter, although extended vastly since Newton's time, is still as in his famous simile, confined to the surrounding shores. Science will have to develop new organs and become amphibious, on this analogy, before transmutation can be an accomplished fact.

Every year brings appreciably nearer the inevitable coming struggle for the possession and control of the primary sources of natural energy. None can foretell the issue and even if the issue were known none but a profound student of modern and contemporaneous history could even faintly outline the consequences, either of success or failure.

Upon this issue, which is purely physical or materialistic and of which the decision rests with the progress of physical science, appears to hang the whole destiny of the race, spiritual, intellectual and aesthetic as well as physical or material.

A race, triumphant and glorying in the million-fold accession of its present recently controlled physical powers, must develop in a manner remote from that of a race struggling on in a world of diminishing reserves of energy, forced out of an opulent civilisation back into its former physical condition of hand to mouth existence upon the daily supply of solar radiation.

*Glasgow, University*

## Note

*Reprinted from *Scientia – International Review of Scientific Synthesis (Via Guastalla 9, Milano, Italy)*, Vol. XI, Annus VI, 1912, pp. 186–202.

# INTRA-ATOMIC CHARGE*

That the intra-atomic charge of an element is determined by its place in the periodic table rather than by its atomic weight, as concluded by A. van den Broek (*Nature*, November 27, p. 372), is strongly supported by the recent generalisation as to the radio-elements and the periodic law. The successive expulsion of one $\alpha$ and two $\beta$ particles in three radio-active changes in any order brings the intra-atomic charge of the element back to its initial value, and the element back to its original place in the table, though its atomic mass is reduced by four units. We have recently obtained something like a direct proof of van den Broek's view that the intra-atomic charge of the nucleus of an atom is not a purely positive charge, as on Rutherford's tentative theory, but is the difference between a positive and a smaller negative charge.

Fajans, in his paper on the periodic law generalisation (*Physikal. Zeitsch.*, 1913, Vol. xiv, p. 131), directed attention to the fact that the changes of chemical nature consequent upon the expulsion of $\alpha$ and $\beta$ particles are precisely of the same kind as in ordinary electrochemical changes of valency. He drew from this the conclusion that radio-active changes must occur in the same region of atomic structure as ordinary chemical changes, rather than with a distinct inner region of structure, or 'nucleus', as hitherto supposed. In my paper on the same generalisation, published immediately after that of Fajans (*Chem. News*, February 28), I laid stress on the absolute identity of chemical properties of different elements occupying the same place in the periodic table.

A simple deduction from this view supplied me with a means of testing the correctness of Fajans's conclusion that radio-changes and chemical changes are concerned with the same region of atomic structure. On my view his conclusion would involve nothing else than that, for example, uranium in its tetravalent uranous compounds must be chemically identical with and non-separable from thorium compounds. For uranium X, formed from uranium I by expulsion of an $\alpha$ particle, is chemically identical with thorium, as also is ionium formed in the same way from uranium II. Uranium X loses two $\beta$ particles and passes back into uranium II, chemically identical with uranium. Uranous salts also

*George B. Kauffman (ed.), Frederick Soddy (1877–1956)*, 19–21.

lose two electrons and pass into the more common hexavalent uranyl compounds. If these electrons come from the same region of the atom uranous salts should be chemically non-separable from thorium salts. But they are not.

There is a strong resemblance in chemical character between uranous and thorium salts, and I asked Mr. Fleck to examine whether they could be separated by chemical methods when mixed, the uranium being kept unchanged throughout in the uranous or tetravalent condition. Mr. Fleck will publish the experiments separately, and I am indebted to him for the result that the two classes of compounds can readily be separated by fractionation methods.

This, I think, amounts to a proof that the electrons expelled as $\beta$ rays come from a nucleus not capable of supplying electrons to or withdrawing them from the ring, though this ring is capable of gaining or losing electrons from the exterior during ordinary electro-chemical changes of valency.

I regard van den Broek's view, that the number representing the net positive charge of the nucleus is the number of the place which the element occupies in the periodic table when all the possible places from hydrogen to uranium are arranged in sequence, as practically proved so far as the relative value of the charge for the members of the end of the sequence, from thallium to uranium, is concerned. We are left uncertain as to the absolute value of the charge, because of the doubt regarding the exact number of rare-earth elements that exist. If we assume that all of these are known, the value for the positive charge of the nucleus of the uranium atom is about 90. Whereas if we make the more doubtful assumption that the periodic table runs regularly, as regards numbers of places, through the rare-earth group, and that between barium and radium, for example, two complete long periods exist, the number is 96. In either case it is appreciably less than 120, the number were the charge equal to one-half the atomic weight, as it would be if the nucleus were made out of $\alpha$ particles only. Six nuclear electrons are known to exist in the uranium atom, which expels in its changes six $\beta$ rays. Were the nucleus made up of $\alpha$ particles there must be thirty or twenty-four respectively nuclear electrons, compared with ninety-six or 102 respectively in the ring. If, as has been suggested, hydrogen is a second component of atomic structure, there must be more than this. But there can be no doubt that there must be some, and that the central charge of the atom on Rutherford's theory cannot be a pure positive charge, but must contain electrons, as van den Broek concludes.

So far as I personally am concerned, this has resulted in a great clarification of my ideas, and it may be helpful to others, though no doubt there is little originality in it. The same algebraic sum of the positive and negative charges in the nucleus, when the arithmetical sum is different, gives what I call 'isotopes' or 'isotopic elements,' because they occupy the same place in the periodic table. They are chemically identical, and save only as regards the relatively few physical properties which depend upon atomic mass directly, physically identical also. Unit changes of this nuclear charge, so reckoned algebraically, give the successive places in the periodic table. For any one 'place,' or any one nuclear charge, more than one number of electrons in the outer-ring system may exist, and in such a case the element exhibits variable valency. But such changes of number, or of valency, concern only the ring and its external environment. There is no in- and out-going of electrons between ring and nucleus.

*Physical Chemistry Laboratory,*
*University of Glasgow*

FREDERICK SODDY

## Note

*Reprinted from *Nature* **92** (4 December 1913): 399–400.

## 'FOREWORD' TO *THE FRUSTRATION OF SCIENCE* (1935)

It is very gratifying to be asked to write a foreword to this book. It deals with what I called long ago the inversion of science, and it is indicative of the growing sense of social responsibility, among some individual scientific men at least, for the world the labours of their order have so largely created. The public expect far more from scientific men in this respect than they have as yet contributed. Individually most of them in this field are still utterly unscientific, and quite as apt as the public themselves to regard original thought on these subjects as socially dangerous and to be suppressed and those who have strayed from the path of 'pure' science in these directions as cranks or impostors. As for the official and professional bodies representing science and medicine, as yet they have hardly emerged from the easy but very questionable attitude that it is no concern of theirs what they are hired out for.

On the other hand, the public must not expect too much. They are apt to forget that in effect, as an entity with power of acting, they hardly exist, until *in extremis* when it is too late. The pioneer and bearer of a new evangel is always up against an inchoate mass, educable only when miserable and, when prosperous, too proud to learn. This much at least of justification can be offered for the doctrine, so utterly the opposite of the truth for the individual, that suffering is the great goad to progress. Unfortunately, scientific powers of inflicting mass suffering are now so powerful that once started they are hardly likely to stop so long as there is anyone left to suffer.

From the beginning to the end of these pages, whether we read of the wilful destruction of the products and productivity of the soil, the aerial destruction of wealth by the thousand million pounds' worth, created by somebody's labour, the embarrassing fecundity of modern technology resulting only in every conceivable form of sabotage, the anomalous position of the conscientious medical practitioner, the refusal of women to bring children into such a world, the development of the art of spreading bacterial infection as a new war technique, or the frank abandonment by modern political movements of the hope of social progress that science renders possible – from the beginning to the end of these pages the reader will find elegant examples of the sort of ruling

*George B. Kauffman (ed.), Frederick Soddy (1877–1956)*, 23–24.

mentality now dominating the world. Bitter, and justifiably so, as many of the critics of science are, surely nothing bitterer could be said of it than this, that its abundance has but enthroned the wastrel. Nor is the solution exactly what one of the contributors rather naïvely suggests, that science should look for a new master. The solution is for the public to acknowledge its real master, and, for its own safety, insist on being ruled not by the reflection of a reflection, but direct by those who are concerned with the creation of its wealth rather than of its debts. It should require that its universities and learned societies should no longer evade their responsibilities and hide under the guise of false humility as the hired servants of the world their work has made possible, but do that for which they are supported in cultured release from routine occupations, and speak the truth though the heavens fall.

FREDERICK SODDY

# PART II

# BACKGROUND

LAWRENCE BADASH

# 1. THE SUICIDAL SUCCESS OF RADIOCHEMISTRY*

*During the few years prior to World War I the science of radiochemistry saw such notable advances that, in effect, it had no more questions to ask, and it was abandoned (except for applications). The group displacement laws offered a traffic pattern for the periodic table, indicating where the daughter products of alpha- and beta-decay should be placed. The concept of isotopy permitted more than one radioelement to exist in a given box of the periodic table. The paths to these great accomplishments were numerous and intertwined. Confirmation by atomic weight measurements was more direct, and equally stirring. Human interest is added to this story of outstanding theoretical and experimental achievement by charges of plagiarism of scientific ideas.*

In his presidential address to the chemistry section of the British Association in 1907, Arthur Smithells pointed to work in radioactivity with wonder, calling it the 'chemistry of phantoms'.[1] Indeed, the transitory nature of the radioelements, coupled with the exceedingly small quantities commonly handled, made many a traditional chemist hesitant to accept these unusual substances as real elements worthy of insertion into the periodic table. Besides, there were too many of them: by 1913 over thirty radioelements were known, but there were no more than twelve boxes in the periodic table in which to house them. Moreover, there was much confusion about radioelements that had different physical properties such as half-life and range of emitted alpha particle, but which could not be separated chemically. Small wonder then that Alexander Russell, the only person who worked with both Ernest Rutherford and Frederick Soddy, recalled the prevalent attitude of chemists as discouraging of interpretative attempts: theirs was, so they claimed, 'an experimental science. No good ever came from pontificating on the ways of Nature from the comfort of an armchair. The laboratory bench, not the sofa, ... was where the truth would be found'.[2]

Yet, these traditional chemists, who kept such a disdainful distance from the subject of radioactivity that the few *radio*chemists in the world worked in far closer contact with physicists, simply were not aware of

*George B. Kauffman (ed.), Frederick Soddy (1877–1956)*, 27–41.

the amount of accumulated experimental data from which theories could be spun. In fact, the data came from an impressively diverse number of scientific specialities.

The main problems, of course, were to identify each radioelement chemically, and to explain the decay-series transitions. With the alpha particle proved to be a charged helium atom, atomic weights could be rather well estimated by calculating the number of alpha-transitions down from uranium, radium, and thorium, whose weights were known. This device served to suggest several chemical identities.

Wet chemical experimentation was a more reliable technique, and from the time of Marie Curie's first encounter with polonium and radium, this was the preferred means of identification. But it was one thing to have radium precipitate together with barium and be separable from it by more delicate processes, and quite another to have radio-thorium and ionium *completely* inseparable from thorium. A whole catalogue of these radioelements with most powerful affinities was assembled by such radiochemists as Otto Hahn, Bertram Boltwood, Herbert McCoy, and Frederick Soddy. When Bruno Keetman, in Berlin, confirmed in 1909 the closeness of ionium and thorium – neither precipitation, nor crystallization, nor sublimation altered the relative amounts of his sample – this chemical closeness had to be regarded as something fundamental.[3]

Other clues came from many sources. At Yale University, Perry Perkins in 1908 measured the molecular weight of radium's gaseous product or emanation by a diffusion experiment, and placed it near radium in the periodic table.[4] At the University of Uppsala, D. Strömholm and The Svedberg applied concepts of isomorphism to the radioelements, implying chemical identity by similar crystalline forms.[5] Friedrich von Lerch in Vienna and Richard Lucas in Leipzig postulated electrochemical relationships between the radioelements.[6] And Georg von Hevesy, doing most of his work in Rutherford's Manchester laboratory, determined the valencies of many radioelements by measuring the diffusion velocities of their ions.[7]

Amidst this generous quantity of data there was one obvious gap: knowledge of the chemical identities of the short-lived beta-emitters. Their very brief existence made wet chemical experimentation difficult, but Soddy, in Glasgow, set Alexander Fleck upon the problem in 1912.

Despite the traditional chemical wisdom not to theorize, the data of radioactivity proved too tempting. Of course, the fundamental explanation of the phenomenon, provided by Rutherford and Soddy in 1902 and 1903, held that one radioelement transforms into another with the emission of radiation. So parent and daughter products in a decay series could be expected to be near one another in the periodic table. Based on their isomorphic investigations, Strömholm and Svedberg saw similarities – corresponding radioelements – in the different decay series, and wondered in print whether these genetic chains extended down the periodic table among the stable elements, the atomic weight of each of which would be an average of its components.[8] Both they and Keetman saw the need to place more than one substance in a given box of the periodic table, and their work helped Soddy in 1911 to gather the courage to use the term 'identical', instead of 'similar', in describing some radioelements.

In that year Soddy published a thin book called *The chemistry of the radio-elements*. In it he proposed a traffic plan for alpha-decay, namely, that the daughter radioelement was two boxes away from the parent in the periodic table, generally in the direction of the lower atomic weight, but possibly also in the opposite direction. The chemistry of the beta-emitting radioelements was too uncertain at the time to characterize their transitions.

In mid-1912, Alexander Russell too tried his hand at theory-construction. As an alumnus of the Glasgow laboratory, he was well informed of Fleck's progress on the short-lived beta-emitters. And now a research student in Rutherford's Manchester laboratory, he was intimately acquainted with Hevesy's electrochemical and valency work, and with the branching experiments first of Kasimir Fajans and later of Ernest Marsden and C. G. Darwin. Based on this latest information, Russell attempted to place the radioelements in boxes of the periodic table. While an improvement over earlier attempts, chemical identification was not complete, and he erred in assigning several products to their groups. Nevertheless, he felt able to offer a complement to Soddy's rule that in alpha-decay there is a group change of two units. In beta-decay, Russell said, there is a group change of one unit, and the change may progress in *either* direction in the periodic table. However, neither Soddy nor Rutherford expressed any interest in this idea.[9]

Towards the end of 1912, Hevesy wrote up his work on valency and sent it to Russell, who improved the English for him. At the same time Hevesy visited Fajans, now in Karlsruhe, and learned that he too was working on a scheme to fit radioelements into the periodic table.

Hevesy thereupon urged Russell to publish his own ideas before they might perhaps be anticipated.[10] It is apparent that as 1912 closed, Soddy, Fajans, Hevesy, and Russell were each aware that at least one other was moving toward resolution of a major problem.

Hevesy published in the *Physikalische Zeitschrift* on 15 January 1913, indicating jumps of two boxes to the left, or an unknown jump, for alpha-decay, and jumps of zero, one, two, or an unknown number of boxes to the right for beta-decay.[11] Obviously, the chemical identity of too many radioelements remained murky.

Russell appeared next in print, at the end of January, in the *Chemical news*. There were some changes, not all of them for the better, from his draft of half a year earlier, but his rules for decay remained unchanged: alpha expulsion meant the daughter product was two groups higher *or* lower; beta expulsion, one group higher *or* lower. And his arrangement of the radioelements showed decay descending from the higher groups of one period, through zero, and then *ascending* through the groups of the next period above.[12]

Neither Hevesy's nor Russell's rules or chemical identifications were correct. But they were close enough to make it apparent that the 'true' interpretation must be something similar. On 15 February 1913, two papers by Fajans appeared that fitted everything magnificently into place. They had been in the hands of the editor of the *Physikalische Zeitschrift* on the last day of the preceding year, so they could not have been influenced by Russell's published paper. They were, in fact, different in approach, being based primarily on electrochemical data, and only partly on radiochemical identities.[13] Possibly they had benefited from Fajans' conversation with Hevesy some months earlier, though Hevesy seems to have been wary of revealing too much information.[14]

In his first paper Fajans organized all the evidence from electrolysis and from dipping metals in solutions that he and other investigators had generated. From these data he saw that an alpha emission left the daughter more electropositive, while a beta emission yielded a daughter more electronegative or nobler. He then applied his rules, and also analogies between the decay series, to radioelements whose chemical natures were as yet untested, assigning them electrochemical relationships to products already identified. In his second paper Fajans used these relationships to place all the radioelements in the periodic table. An electropositive transformation meant movement from right to left in a horizontal period; Fajans accepted Soddy's conclusion that the jump was of two boxes. Conversely, an electronegative transition indicated left-to-right movement, and here Fajans decided it was of just *one* box. Another crucial point was his statement that the decay series moves from higher to lower groups in *both* periods, and the transition between periods occurs from emanation (group 0) to an A product (group VI). The sequence could best be visualized if the two-dimensional periodic table were rolled into a helix. Not all was perfect, for Fajans' electrochemical logic forced him to postulate the existence of a few imaginary radioelements, but his rules, and chemical identities, and prediction of the existence of uranium-$X_2$ (an isotope of protactinium, which he soon found) were seen to resolve so well the uncertainties of radiochemistry that the subject soon had no future.

Less than two weeks after this, on 28 February 1913, a paper by Soddy appeared in the *Chemical news* containing essentially the same information. Like Fajans, he placed all the known radioelements into the periodic table; unlike Fajans, his approach was through chemical identities, not electrochemistry, so he saw no need to suggest the existence of Fajans' extra radioelements. The postulation of uranium-$X_2$, however, was seen to be necessary, and like Fajans, Soddy correctly placed it below tantalum in group five. Again like Fajans, he emphasized that radioelements located in the same box would be chemically inseparable, even though they belonged to different decay series, and he too found the atomic weight of inactive end-product lead a promising key to the confirmation of these identical-but-different substances.[15]

The puzzling thing about this paper is why it ever was published. It can hardly be called a case of simultaneous and independent discovery, for Soddy actually acknowledged in the text of his article that he had read Fajans' paper and accepted its ideas. The stated reason for publishing was that his paper was already drafted when Fajans' came to hand. Presumably Soddy also felt that Fleck, who was furnishing so much of the experimental evidence that identified the short-lived radioelements chemically, was working in his laboratory at his own suggestion, and Soddy himself had been intimately connected with the subject. And of course his paper approached the question from a point of view different from that of Fajans, namely chemical identity. Had Soddy offered this as confirmation of Fajans' ideas with different evidence, and even as an extension of the consequences to include the probable lack of spectroscopic distinctions between substances in the same box, this would have been acceptable. Had he criticized the errors of Russell and Fajans, which instead he rejected tacitly, this too would have been acceptable. But presenting it as an independent discovery was a surprising violation of scientific etiquette. Possibly it was even more.

On 13 February 1913 two days before Fajans' papers were published, Fleck sent a letter to the *Chemical news* commenting on Russell's earlier paper. Russell had indicated that thorium-D (now Th-C″) should be analogous either to lead or to mercury. Fleck now described experiments proving it could not be the former. 'It thus appears', he concluded, 'to be much more nearly allied to the ... mercury sub-group than to the ... lead sub-group'.[16]

Despite Fleck's false conclusion, when Soddy wrote his paper five days later, on 18 February, he correctly identified thorium-D not with mercury, but with thallium, as Fajans had done. Fleck *may* have immediately repudiated his published conclusion, but we have no indication of this, and moreover, he denied to me personally that he had quickly changed his mind.[17] Soddy *may* have had an inspiration to place thorium-D in group III, but his earlier behaviour indicated reluctance to speculate beyond Fleck's results. In fact, in his Nobel Prize address, Soddy was later to remark that 'till then I had scrupulously avoided even thinking about [Fleck's] results until they were completed, for fear of giving an unconscious bias to the work in progress in the laboratory'.[18]

With thorium-D misidentified, the group displacement laws were inconsistent, and the scheme little better than Russell's. With thorium-D correctly placed in the thallium box, the displacement laws were consistent. It is likely that Fajans' paper caused everything to fall into place in Soddy's mind, and Soddy, being so close to the material, could not later sort out what he had borrowed from another.

Without question, Soddy was a giant among those working in radioactivity. The transformation theory of radioactivity (with Rutherford), the concept of isotopy, and the initial formulation of the alpha decay rule were major achievements. But the final, correct interpretation of the group displacement laws belongs to Kasimir Fajans alone.

Fajans, moreover, provided a powerful confirmation of these new ideas. A scientific theory that explains existing knowledge is worthwhile; one that is pregnant with the unknown is exciting as well. Such predictions furnish the means by which the theory may be tested. In the case of the group displacement laws verification came by several means. With his student, Oswald Göhring, Fajans experimentally discovered the predicted uranium-$X^2$. This was especially interesting because no other 'eka-tantalum' was known. At a time when isotopes were recognized for virtually all the heaviest elements, here was a body which filled its box (now element 91) in the periodic table alone. Because of its brief half-life (1.1 minutes), they named the new radioelement 'brevium'. It was not, as first suggested, the long-sought parent of actinium, but when that substance eventually was discovered and found to be a long-lived isotope of brevium, Fajans agreed that its name, 'protactinium', should thenceforth designate all varieties of the element.[19] Fleck was also busy in the spring of 1913, confirming by chemical test several more radioelements whose identities Fajans had predicted. Such work reinforced confidence in the accuracy of the group displacement laws.[20]

But note that these confirmations related essentially to the displacement laws alone. Welcome as they were, they did not remove all doubts that isotopes existed only in the fertile imaginations of a few radiochemists and other exotic specialists. The iron-clad proof required, as clearly seen by Fajans and Soddy, was a series of comparative atomic weight determinations of isotopic elements. Unfortunately, scarcely any radioelement was obtainable both in isotopic purity and in quantity sufficient for its atomic weight to be measured. This was part of the problem surrounding the question whether they deserved to be called elements at all. There was, however, the inactive lead end-product of the uranium decay series, available in satisfactory amounts, and the likelihood that the thorium chain also ended in lead. Ordinary lead had an established atomic weight of 207.15. Taking that of radium as 226.0 and deducting for five alpha particle expulsions of four units each, the uranium-series lead should weigh 206.0 (or 206.5 if calculated as eight alphas down from uranium at 238.5). Similarly, six alpha transitions from thorium at 232.4 indicated that thorium-series lead should have an atomic weight of 208.4.

Before this time atomic weight experts had had little to do with radioactivity. Indeed, these painstaking precision chemists may be viewed as logical opponents to ideas of isotopy, for had they not devoted their whole careers to ascertaining the very best *single* value for each element? Still, vested interest or not, the experimental test was one within their capabilities, and little further stimulus was needed. Fajans and Soddy each tried to make the atomic weight determinations themselves. Fajans soon realized that the task required such accuracy that only the results of recognized experts would be widely accepted. Soddy persisted in his efforts, only to have his results viewed sceptically in some quarters.[21]

The recognized master that Fajans chose for this work was the world's leading atomic weight specialist, Theodore W. Richards of Harvard University. Infinite care was the hallmark of his trade, for impurities would render experimental results valueless. The action of acids upon glass, the occlusion of gases by other substances, the need to prevent dried substances from absorbing moisture before being weighed – all had to be meticulously controlled. Fajans arranged for Max Lembert, a Karlsruhe student just beginning doctoral research, to work in Richards's laboratory. When Lembert arrived in Cambridge, Massachusetts, in the autumn of 1913, to determine the atomic weight of lead from radioactive sources, Richards was sceptical though enthusiastic, for Sir William Ramsay, at Soddy's request, had also recently urged the problem on him.[22]

Control experiments on ordinary lead were carried out and yielded the accepted atomic weight, 207.15. The weights of the radioactive samples tested, however, were strikingly lower. Fajans had provided lead from Colorado carnotite, which he obtained from Friedrich Giesel in Braunschweig, and this gave 206.59; another sample he sent was extracted from Joachimsthal pitchblende (206.57). Ramsay furnished residues from Cornwall pitchblende (206.86). Boltwood of Yale provided lead from Ceylonese thorianite, as did H. S. Miner, chief chemist of the Welsbach Light Company in New Jersey (206.82). This mineral contained some three times as much thorium as uranium, but since the latter decays about three times as fast as the former, the amounts of lead from each series were expected to be roughly equal. Ellen Gleditsch, of Oslo but then working with Boltwood, gave Richards a particularly pure and hence valuable lead from North Carolina uraninite (206.40).[23]

Experimental comparisons of other elements, from widely scattered sources on the globe and even from meteorites, had never before yielded discordant results. In this case, Richards, who presumably composed the paper he published with Lembert, studded it with the words 'amazing', 'revolutionary', 'extraordinary', and 'striking'. There was no doubt that radioactive lead was decidedly lighter than ordinary lead. Spectroscopic examination, moreover, revealed no differences whatever. Yet Richards was too much a chemist of the old school to drop his reservations entirely and proclaim aloud the truth of isotopy. Instead, he spoke of an 'admixture' having reactions similar if not identical to lead, though of a lower atomic weight. The admixture might have the same spectrum as lead, or have no spectrum in the ultraviolet region examined, or have its spectrum masked by the greater quantities of ordinary lead. Another possibility, he conceded, with reference to the work of Fajans and Soddy, was that ordinary lead was a 'medley' of substances of different origin but similar properties, in varying proportions.

These results were reported in late May and June 1914. Slightly earlier Soddy had announced the result of his own work with Ceylonese thorite: a value of 208.4, which was significantly higher than ordinary lead's 207.15, and only slightly higher than the theoretical weight deduced for a mineral of this composition.[24] Maurice Curie, nephew of Marie, was the next to offer similar evidence. In early June the Paris newspaper *Le Matin* headlined a front-page story 'Il y a plusieurs espèces de plomb', describing Curie's report to the Académie des sciences. He had examined the lead from three uranium minerals, one thorium mineral, and common galena, finding lower and higher atomic weights as expected.[25]

The most convincing confirmation of the Harvard work came, however, from another atomic weight expert. Just a few years earlier, Otto Hönigschmid had learned the techniques in Richards's laboratory.[26] Now he was one of the bright young men in the field, making a name for himself in Prague and Vienna with measurements of radium and uranium. With his student, Stephanie Horovitz, Hönigschmid determined the atomic weight of lead from Joachimsthal pitchblende, obtaining a value of 206.736. This was published a week after Maurice Curie's paper, although the results had been presented on 23 May 1914 at a congress of the Bunsen Gesellschaft in Leipzig, at which meeting Fajans also reported upon the work of Richards and Lembert.[27] Thus within about a single month the existence of isotopes was confirmed independently in four laboratories in four countries.

Some, such as Samuel Glasstone, recall the astonishment – 'almost consternation' – of the period. But the evidence was so strong that prejudices were swept aside.[28] Others, such as James B. Conant, had been earlier convinced by the radioactivity arguments, and awaited the Richards and Lembert results with equanimity.[29] Regardless of preconceived notions, it was a time of intellectual excitement, not entirely overcast by the outbreak of World War I.

In the next few years Richards and Hönigschmid and their students amassed further evidence of different atomic weights for lead. Soon it was generally accepted that the thorium series must end in lead, and lead only. The possibility that branches in the decay series might *not* rejoin was disposed of by attrition: no evidence in its favour was produced. The parent of actinium, protactinium, was discovered towards the end of the war, but whether this series was connected to the uranium series, or descended from a yet unknown isotope of uranium, was one of the few problems in radiochemistry to remain for the 1920s, when F. W. Aston's mass spectrographic work showed the existence of uranium-235.[30]

Other work that followed as a consequence of the group displacement laws and the concept of isotopy included extensive studies of spectra, thermoelectric effects, refractive indices, solubilities, melting points, atomic volumes and atomic densities. As Soddy remarked about isotopes in his Nobel Lecture, 'put colloquially, their atoms have identical outsides but different insides'.[31] These studies, therefore, were motivated by the understanding that certain atomic properties depended upon the 'outside' electron configuration, and others were due to the 'inside' mass of the nucleus. Small differences in line spectra, and larger variations in band spectra of isotopes were discovered eventually. Similarly, slight differences in chemical reaction rates affected by mass also were found. Yet just as radioactivity was something of a hybrid between physics and chemistry, so too was this new field; it was sometimes called 'modern chemistry', though more often 'atomic physics'.

This evolution of one science into another as old problems were solved and new ones arose was made easier by the touchstone of traditional chemistry. What had been radical ideas about the nature of elements but a few years before could be accepted, at least provisionally, because of the classical, wet-chemistry atomic weight evidence in favour of the hypotheses of Fajans and Soddy. Until 1919 it was the only technique available for atomic weight determinations. In that year Aston developed his mass spectrograph, which provided precision measurements of atomic masses through the deflection of ions in a magnetic field, and clearly showed different masses for a given element. The method satisfied physicists, but chemists felt it yielded atomic weights by a mere process of calculation, and was thus little better in principle than subtracting four units for each alpha particle emission.

Chemists did not desire such arithmetic; they much preferred laboratory measurement.[32] The initiation and acceptance of the many investigations of isotopic properties had already relied substantially upon the imprimatur given by the work by traditional methods of such chemists as Richards and Hönigschmid. Increasingly, chemists would come to respect the value of the mass spectrograph, but during the second decade of the century classical chemistry was a more-than-sufficient bedrock for the new ideas.

Still, for those who looked closely at the meaning of these concepts, not all questions were resolved. For several years following the introduction of the idea of isotopes a subdued controversy raged over their interpretation. Aston's work made it clear that a multiplicity of forms for a body with a given atomic number was a general phenomenon, not restricted to radioelements. Fajans, seeking to disturb the age-old definition of 'element' as little as possible, looked upon each new isotope as a different element. Uranium-238 and U-234, therefore, were individual elements because of differences in such properties as mass and half-life. Most other chemists and physicists, however, regarded chemical combination as the predominant yardstick in the definition of an element, and were content to accept isotopes as varieties of the same element. This new interpretation, which removed atomic weight as an indicator of the identity of an element, was smoothed by H. G. J. Moseley's work showing the significance of atomic number. By 1922, Fajans accepted the popular view.[33]

Another problem of interpretation that arose concerned precisely what occurred in the atom during radioactive decay. As early as February 1913, Fajans declared it to be a surface phenomenon, because changes in the chemical nature of the atom due to alpha and beta particle expulsion were of the same type as resulted from electrochemical changes of valency.[34] It was a view difficult to reconcile with Rutherford's model of a nuclear atom though, for the only 'occupants' of the surface region, where chemical reactions occurred, were orbital electrons. These might furnish the beta particles, yet where else but the nucleus could the alphas arise? There was, in fact, strong circumstantial evidence that radioactivity was solely a nuclear phenomenon. Not only was it logical to expect atomic-size alphas to come from the material nucleus, but this core was also seen as composed of differing numbers of positive and negative particles, with a net positive sum equal to the element's place in the periodic table. The ideas of the Dutch physicist

A. van den Broek on nuclear constituents, supported by the importance given to the sequential arrangement of elements by the displacement laws, by Niels Bohr's relegation of radioactivity to the nucleus, and by Moseley's arrangement of the elements according to atomic number, all made it appear that beta particles were nuclear electrons, those negative bodies that allowed hydrogen nuclei to make up the atom's measured atomic weight, but reduced their positive charge to the appropriate atomic number. By the end of 1913, Soddy offered a further argument against viewing radioactivity as a surface phenomenon. Uranium in its uranous form is tetravalent, he pointed out, but in its more common uranyl form is hexavalent, a loss of two orbital electrons. Similarly, uranium-X, upon the emission of two beta particles, is transformed into uranium-II. If electron loss and beta loss were identical processes, Soddy argued, uranous salts and compounds of uranium-X would be chemically non-separable. But uranium-X is an isotope of thorium, and clearly separable from uranium.[35]

Beyond such matters of theoretical significance there was an abundant range of experimental investigations into the effects of radioactivity. Yet, like the work mentioned earlier, this too was a departure from the main line of fundamental radioactivity studies. While both physicists and chemists participated, a large portion of it was conventional chemistry, merely using radioactive materials. In a few instances radioactivity also found applications, the most exciting and significant being the illumination of the nature of precipitation and adsorption by Fajans, Hahn, and F. A. Paneth, who used such small quantities of matter that only radioactive means could serve for their detection. But the leading pioneer in tracer techniques was Hevesy, who extended the application of such indicators to biological and medical systems.

These uses of radioactivity were the wave of the future; the past had yielded the basic science of radiochemistry upon which such applications rested. By the early 1920s, then, almost all the radioelements were known, their sequences in the decay series determined, fundamental understandings such as the disintegration theory of radioactivity, the group displacement laws, and isotopy enunciated and verified, and innumerable quantitative relationships established. Chemical studies of the radioelements effectively ended, for the questions this science asked had been satisfactorily answered. The opportunities for basic research in radiochemistry were no more; radiochemistry was suicidally successful. Not until the mid-1930s, following the discoveries of artificial radioactivity and the neutron, was the subject resurrected, then to be called nuclear chemistry.

*Department of History*
*University of California, Santa Barbara*
*Santa Barbara, CA 93106, U.S.A.*

## Notes

* Reprinted from the *British Journal for the History of Science* **12**: 42 (November 1979): 245–56.

[1] A. Smithells, 'Presidential Address to Section B', *Nature*, 8 August 1907, **76**, 352–357.

[2] A. S. Russell, 'Lord Rutherford: Manchester, 1907–19: A Partial Portrait', *Proceedings of the Physical Society of London*, 1951, **64**, 224–225.

[3] *Ibid.*

[4] P. B. Perkins, 'A Determination of the Molecular Weight of Radium Emanation by the Comparison of its Rate of Diffusion with Mercury Vapor', *American Journal of Science*, 1908, **25**, 461–473.

[5] D. Strömholm and T. Svedberg, 'Untersuchungen über die Chemie der radioaktiven Grundstoffe', *Zeitschrift für anorganische Chemie*, 1909, **61**, 338–346; 1909, **63**, 197–206.

[6] F. von Lerch, 'Die physikalischen und chemischen Eigenschaften der Umwandlungsprodukte des Thoriums', *Jahrbuch der Radioaktivität und Elektronik*, 1905, **2**, 461–475. R. Lucas, 'Über das elektrochemische Verhalten der radioaktiven Elemente', *Physikalische Zeitschrift*, 1906, **1**, 340–342.

[7] G. von Hevesy, 'The Electrochemistry of Radioactive Bodies', *Philosophical Magazine*, 1912, **23**, 628–646.

[8] D. Strömholm and T. Svedberg, *op. cit.* (5).

[9] A. S. Russell, letter to E. Rutherford, 14 September 1912, Rutherford collection, Cambridge University Library (hereafter cited as RCC). A. S. Russell, conversation with the author, 20 April 1970. James Chadwick, conversation with the author, 19 February 1970.

[10] G. von Hevesy, letter to E. Rutherford, 7 December 1912, RCC.

[11] G. von Hevesy, 'The Valency of the Radioelements', *Philosophical Magazine*, 1913, **25**, 390–414; 'Die Valenz der Radioelemente', *Physikalische Zeitschrift*, 1913, **14**, 49–62, Hevesy letters to E. Rutherford, 7 December 1912 and 3 January 1913, RCC.

[12] A. S. Russell, 'The Periodic System and the Radio-Elements', *Chemical News*, 31 January 1913, **107**, 49–52.

[13] K. Fajans, 'Über eine Beziehung zwischen der Art einer radioaktiven Umwandlung und dem elektrochemischen Verhalten der betreffenden Radioelemente', *Physikalische Zeitschrift*, 1913, **14**, 131–136; 'Die Stellung der Radioelemente im periodischen System', *ibid.*, pp. 136–142.

[14] G. von Hevesy, letter to A. S. Eve, 28 December 1937, RCC.

[15] F. Soddy, 'The Radio-Elements and the Periodic Law', *Chemical News*, 28 February 1913, **107**, 97–99.
[16] A. Fleck, 'The Periodic System and the Radio-Elements', *Chemical News*, 21 February 1913, **107**, 95.
[17] A. Fleck, conversation with the author, 25 January 1966.
[18] F. Soddy, 'The Origins of the Conception of Isotopes', *Les Prix Nobel en 1921–1922*, Stockholm, 1923, p. 17.
[19] K. Fajans and O. Göhring, 'Über die komplexe Natur des Ur X', *Naturwissenschaften*, 4 April 1913, **14**, 339; 'Über das Uran $X_2$ – das neue Element der Uranreihe', *Physikalische Zeitschrift*, 1913, **14**, 877–884. K. Fajans, 'Discovery and Naming of the Isotopes of the Element 91', remarks prepared for the Third International Protactinium Conference, 14 April 1969.
[20] A. Fleck, 'The Chemical Nature of some Radioactive Disintegration Products. Part II', *Chemical News*, 6 June 1913, **107**, 273; same title, *Journal of the Chemical Society*, 1913, **103**, 1052–1061. The Fajans-Soddy priority question has been discussed in much the same fashion as here by Jozef Hurwic, 'Kasimir Fajans (1887–1975); souvenirs sur sa vie et ses activités', *L'Actualité chimique*, January, 1976, pp. 28–32. For more extensive information on the entire subject, see L. Badash, *Radioactivity in America: Growth and Decay of a Science*, Baltimore, 1979.
[21] K. Fajans, conversation with the author, 16–17 July 1966. K. Fajans, letter to James B. Conant, 30 October 1969, Fajans collection, University of Michigan Library. T. W. Richards, letter to B. Boltwood, 10 January 1917, Boltwood collection, Yale University Library.
[22] K. Fajans, *ibid.* See also, T. W. Richards, letter to K. Fajans, 8 June 1914, Fajans collection, University of Michigan Library.
[23] T. W. Richards and M. Lembert, 'The Atomic Weight of Lead of Radioactive Origin', *Science*, 5 June 1914, **39**, 831–832; same title, *Journal of the American Chemical Society*, 1914, **36**, 1329–1344. T. W. Richards, letter to B. Boltwood, 30 April 1914; and reply, 1 May 1914, Boltwood collection, Yale University Library.
[24] F. Soddy and H. Hyman, 'The Atomic Weight of Lead from Ceylon Thorite', *Journal of the Chemical Society*, 1914, **105**, 1402–1408.
[25] *Le Matin*, 9 June 1914. Maurice Curie, 'Sur les écarts de poids atomiques obtenus avec le plomb provenant de divers minéraux', *Comptes Rendus*, 8 June 1914, **158**, 1676–1679.
[26] T. W. Richards, letters to O. Hönigschmid, 17 December 1912 and 24 February 1913, Harvard University Archives.
[27] O. Hönigschmid and S. Horovitz, 'Sur le poids atomique du plomb de la pechblende', *Comptes rendus*, 15 June 1914, **158**, 1795–1798. F. Soddy, 'Radioactivity', *Annual Reports on the Progress of Chemistry* (Chemical Society of London), 1914, **11**, 266–269.
[28] S. Glasstone, letter to the author, 5 September 1966.
[29] James B. Conant, 'Theodore William Richards and the Periodic Table', *Science*, 24 April 1970, **168**, 425–428.
[30] F. W. Aston, 'The Mass-Spectrum of Uranium Lead and the Atomic Weight of Protactinium', *Nature*, 2 March 1929, **123**, 313. E. Rutherford, 'Origin of Actinium and Age of the Earth', *ibid.*, pp. 313–314.
[31] F. Soddy, *op. cit.* (18), p. 2.
[32] A. S. Russell, conversation with the author, 20 April 1970.

[33] K. Fajans, *Radioactivity*, London, 1923, p. xii.
[34] K. Fajans, 'Die radioaktiven Umwandlungen und die Valenzfrage vom Standpunkte der Struktur der Atome', *Verhandlungen der Deutschen physikalischen Gesellschaft*, 1913, **15**, 240–259.
[35] F. Soddy, 'Intra-atomic Charge', *Nature*, 4 December 1913, **92**, 399–400.

S. B. SINCLAIR

# 2. RADIOACTIVITY AND ITS NINETEENTH-CENTURY BACKGROUND

*One can begin to view the origins of radioactivity in a perspective broader than the merely experimental by considering aspects of the preceding period. In explaining such phenomena as electrical and magnetic actions, optical properties of matter and the behaviour of light, and the nature of chemical matter, connexions between all these were sought. It is pertinent to enquire whether the nineteenth-century unifying ideals relating matter, ether, and electricity were influential in the union of physics and chemistry that was radioactivity. In doing so, one may discern threads and themes of ideas and speculations, more or less closely linked to experimental facts and programmes. Our three threads (chemical, physical, and chemical physics) which converge towards the watershed of radioactivity are linked to three which emerge beyond it (cosmical, Proutian, and universal). These differ in importance but their influence upon the early workers on radioactivity, such as Marie Curie and Ernest Rutherford, is detectable. The case of Frederick Soddy is more debatable; one views with interest his contemporary claim that the transmutation theory of radioactivity was entirely independent of electronic theories of matter.*

I wish to take a rather broad view of the early history of radioactivity in the hope that some light may be thrown upon the conceptions and expressions of those involved in the subject in its earliest stages. Now it is certainly valid to place the origins of the novel theory of radioactive atomic transmutation among the great experimental studies of the last years of the nineteenth century: Röntgen- or x-rays (1895–6); the inert gases (1894–8); the Zeeman effect (1897); the electron or corpuscle (1897); and Becquerel- or uranium-rays (1896) form an exciting picture of discovery. However, the view from a somewhat more distant perspective can reveal aspects of the subject which may not only contribute to an understanding of the science itself, but may also allow some comment upon the striking attitudes which Frederick Soddy adopted after his chemical contribution to the transformation theory of radioactivity.

Looking to the preceding period, to the nineteenth-century back-

*George B. Kauffman (ed.), Frederick Soddy (1877–1956)*, 43–53.

ground, it is possible to discern continuous threads of scientific development, three at least in number. These I intend briefly to outline, bringing them in each case to the point of contact with radioactivity. The first we shall take up can be loosely described as the *chemical thread*. During the last quarter of the nineteenth century a persistent line of chemical speculations considered the unity and complexity of the chemical atoms. They insisted on the point, supported in part by atomic weights, that all of the elements are complex, being composed of a single material; or perhaps two materials; or at least that some heavy elements are composed of lighter ones. In these later developments of Prout's hypothesis, candidates such as hydrogen or the postulated ultra-light spectroscopic sun-element 'helium', or even the ether of space were proposed for the basic materials. The question of the mechanism by which the many were formed from the one was also raised. It was answered by the suggestion that the patterns and families of the comparatively recently established periodic table (1860.f) had arisen by a form of evolution. In the words of one chemist, William Crookes, in 1886:

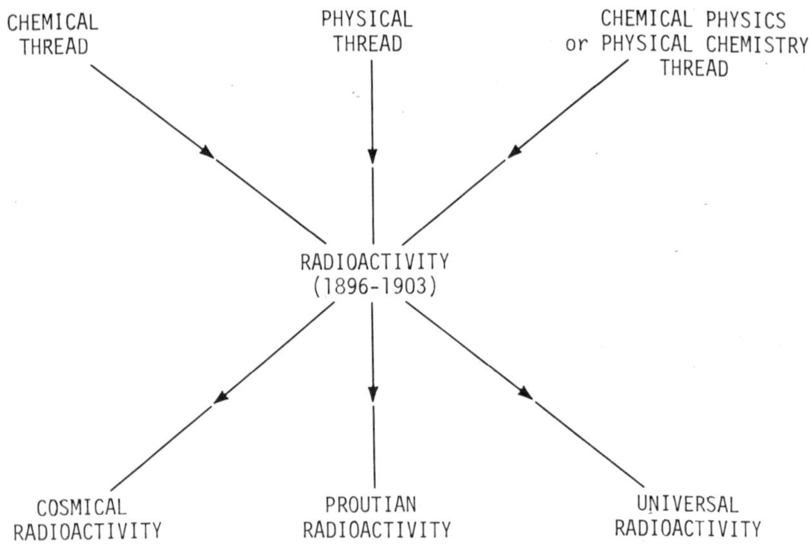


Fig. 1. Radioactivity: threads through the watershed.

It plainly appears that the notion – not necessarily of the decomposability, but at any rate of the complexity of our supposed elements – is, so to speak, in the air of science, waiting to take a further and more definite development. It is important to keep before men's minds the idea of the genesis of the elements; this gives some form to our conceptions, and accustoms the mind to look for some physical production of atoms. It is still more important, too, to keep in view the great probability that there exist in Nature laboratories where atoms are formed and laboratories where atoms cease to be. We are on the track and are not daunted....[1]

Undaunted, and even stimulated by his struggles with the rare earths, Crookes proposed an entire evolutionary scheme of the periodic table. An oscillating cosmic force or pendulum swings back and forth as the original *protyle* condenses and deposits successively the various elements, of which only the most stable have survived. Atomic weight relationships, diamagnetism and paramagnetism, chemical and valency patterns he accommodated into the system. In his opinion, however, 'The epoch of elemental development is decidedly over', and all is now frozen. As for the dissociation of the present elements, in the opinion of the chemists 'The highest temperatures and the most powerful electric currents at our disposal have been tried, and tried in vain'.[2] Not so for the astrophysicist cum astrochemist Norman Lockyer, a figure most renowned for taking this debatable step. In works such as *The Chemistry of the Sun* (1887), and *Inorganic Evolution as studied by Spectrum Analysis* (1900), he argued that the similarities and coincidences between various metallic spectra and between these and stellar and solar spectra indicated that both evolutionary condensation and actual dissociation of some elements occurred in nature and in the laboratory.

In 1898, reviewing studies on radioactivity after her first year's work on the subject, Mme. Curie put forward several speculations on the source of the continually radiated energy, not yet the enormous problem that it was to be. One was that indeed:

The available energy of radioactive substances constantly diminishes. One may for example connect radioactivity with Crookes' theory of the evolution of the elements, attributing radioactivity to elements of high atomic weight which would have been formed last, and whose evolution would not yet be complete.[3]

According to this view, as we see, uranium, thorium, and presumably polonium were still in the process of formation from 'protyle' or primary matter. Though this does not seem to be the hypothesis she favoured, nor was it accepted by Crookes himself at the time, the fact that his idea of more than a decade earlier could be brought up in this way shows

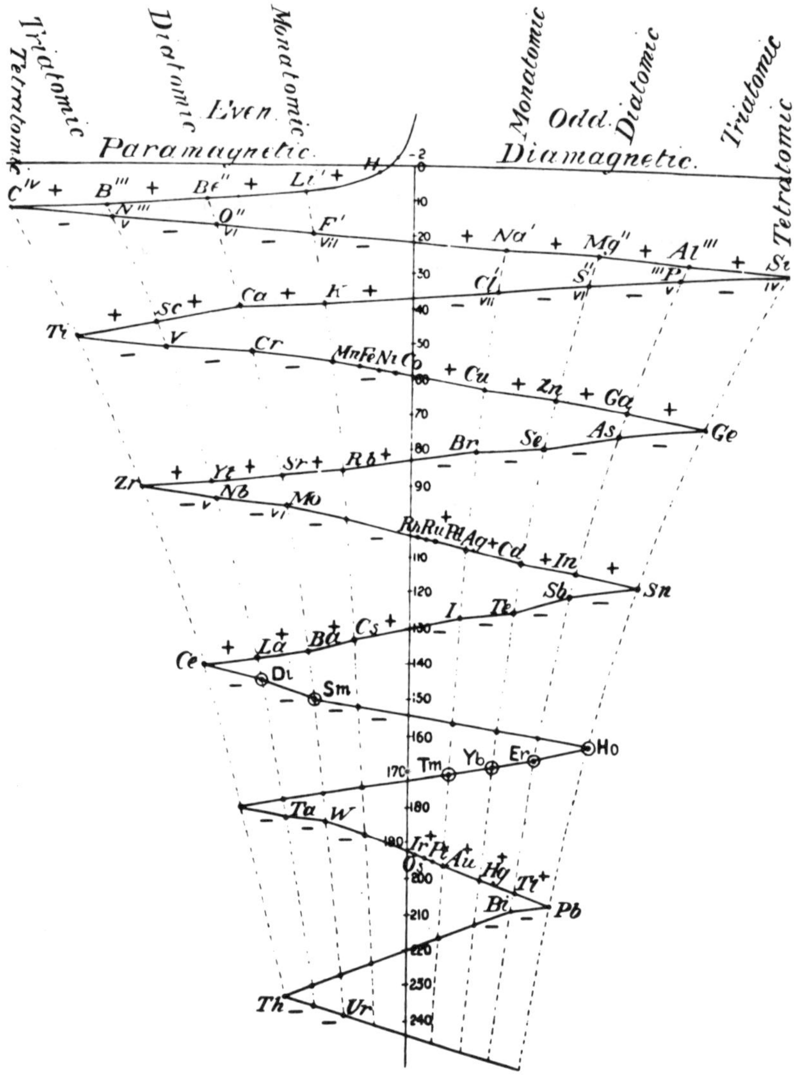


Fig. 2. William Crookes' evolutionary periodic table (1886).

that it was still current at the end of the nineteenth century. In fact, it turned out to be nearer the mark than either had supposed.

A second thread of more distinctly *physical* theories, quite strongly concerned with light as a main theme, again with a unifying aspect, can be traced through to the end of the nineteenth century. A host of optical phenomena such as interference, dispersion, aberration, reflection, polarisation, etc. were explained in terms of an *ether* or *aether* of space as the medium which carried the light waves. Explanations of the experimentally multiplying phenomena were developed by means of theoretical elastic-solid ethers, then moving fluid ethers, to which Newtonian mechanics was mathematically applied. Finally, Maxwell developed a theory of radiation which tended to unite optical and electromagnetic phenomena as manifestations of the same ether, starting from an electrical analogy of the dielectric medium as the basis. Hertz' clear experimental demonstration (about 1888) of the wave-like character of electrical disturbances helped to force a way against the conventional objections. According to a reviewing physicist in 1885, the main problem was the difficulty of 'realising mechanically' the nature of the fundamental 'electric displacement' in an unspecified 'medium':

> The main difficulty of the theory, and the one which stands most in the way of its general acceptance, is the difficulty of forming a clear physical picture of what electric displacement is.[4]

An alternative or further line of development of the electromagnetic theory of light tended to combine a solution of this problem with that of understanding the *discontinuous* aspects of matter and electricity. This can be seen as a second strand of the physical thread. Thus Lorentz developed a *theory of electrons* (1892) which provide electromagnetic optical explanations in terms of ponderable matter and a static ether connected via electrically charged particles, to which the usual principles of Newtonian dynamics and of energy were applied. However, the unifying attitude belongs more to the British school, where, for example, we find Larmor claiming that:

> the electrical development of aether-theory ... involves the explanation of the atomic constitution of matter itself: matter must be constituted of isolated portions ... in and belonging to the aether.[5]

Such isolated portions he conceived as varieties of vortices in a fluid ether, which will concern us later. The significance of the physical line for our present purpose is, firstly, that the study of ether waves provided the subject of some of Rutherford's earliest publications which themselves link with radioactivity in several ways. Experimentally, he turned from a study of one form of penetrating radiation, Hertzian waves, to another, x-rays, and thence to uranium rays; that was in 1897. And theoretically he compared the decay of radioactive rays (from thorium emanation) with the decay of waves from an oscillating electrical circuit: it is of interest that he was preparing a postgraduate course on Electric Waves and Oscillations at the time, in 1899. And the second strand of the physical thread led via Larmor's or Lorentz' electronic theories of the optical behaviour of matter to Zeeman's discovery of the magnetic influence on sources emitting light.[6] The measurable widening of the sodium spectrum doublet by this means gave an immediate experimental boost to J. J. Thomson's conception of a complex chemical atom composed of corpuscles or electrons.[7] *Most* of those concerned with radioactivity saw such a theory as highly relevant to the subject.

This leads us to a third main thread of the nineteenth century which in part unites the previously outlined 'chemical' and 'physical' areas. This may be placed under a heading *chemical physics* or *physical chemistry* in which a rather thin line of physicists attempted to apply their science to chemistry; this was complemented by a few chemists thinking in terms of physical theory. The study of the electrical phenomena exhibited by chemical substances, especially in the gaseous state, including cathode rays (from Faraday onwards) is a case in point as exemplified by the work of the chemist William Crookes and the physicist J. J. Thomson. Now one physicist, Oliver Lodge, reporting on one such area 'Electrolysis' (to the British Association) in 1885, though 'convinced of its immense importance', considered that it possessed 'the somewhat *repulsive* character attaching to any borderland branch of science – in this case not wholly physics nor wholly chemistry'.[8] But others did not take quite this view. Thomson believed that his experiments of 1894 on 'The Connection between Chemical Combination and the Discharge of Electricity through Gases':

> give hopes that the study of the passages of electricity through gases may be the means of throwing light on the mechanism of chemical combination. The work of chemists and physicists may be compared to that of two sets of engineers boring a tunnel from opposite ends – they have not met yet, but they have got so near together that they can hear the sounds of each other's works and appreciate the importance of each other's advances.[9]

Indeed, it is notable that many of the older scientists who had sufficient interest to publish on radioactivity in the first years of the twentieth century were those studying such border areas in the nineteenth. Among chemists were H. E. Armstrong, Crookes, Ostwald, and Mendeleef; physicists include Becquerel, Kelvin, Lodge, G. J. Stoney, A. Schuster, and Thomson himself.

Thompson is one of the strongest figures in this thread of chemical physics. His attempts to explain chemical matter in terms of the ether of physics date back to the early 1880s in his *Treatise on the Motion of Vortex Rings*.[10] The vortex-ring construction of chemical atoms from a perfect fluid ether owed something to an earlier analogy with smoke rings, which can also help us to picture the idea. Vortex rings theoretically had the required stability for chemical atoms, and there was the undeveloped possibility that their vibrations could represent internal atomic energy, which was promising both for thermal and spectroscopic properties. Chemical bonding, with its impermanent nature, was seen as analogous to an observed temporary association of two smoke rings under certain conditions, whereby each would in turn pass through the other, alternately expanding and contracting. In this way Thomson developed a full kinematical-mathematical treatment both of chemical combination and of the pressure, temperature, and diffusion laws of gases. He considered this to be a more fundamental approach than the ordinary 'solid particle' theory which required the assumption of special forces between atoms. As for the structure of individual atoms, only hydrogen might have the single ring; the other atoms were built up of indestructibly assembled vortex fibres twisted like the strands of a rope and joined into rings: 'the strength[11] of the ring and the volume of liquid composing it remain for ever unaltered' and rings 'will retain for ever the same kind of be-knottedness or linking'. The atoms were thus highly complex, but nonetheless unbreakable. Again, the unifying attitude comes out in Thomson's attempt at this time to relate chemical stability to the dielectric strengths of gases:

> Let us now suppose that we have a quantity of gas in an electric field. We shall suppose, as the most general assumption we can make, that the electric field consists of a distribution of velocity in the medium whose vortex-motion constitutes the atoms of the gas.[12]

Theoretical developments ran into difficulties and turned heavily towards the experimental during the following decade;[13] but the hopes of chemical physics emerged triumphant in Thomson's classical paper of 1897 on 'Cathode Rays',[14] in which he announced his discovery of the electron, his 'corpuscle'; he seems indeed to have viewed this as the long-sought prime matter of the chemists. Again, he put forward his attempts to explain chemical matter in physical terms, if newer terms. The atom is still complex but is now divisible to a small extent. The arrangements of the several thousand corpuscles within each atom are to correlate with the patterns and relationships exhibited externally in the chemical periodic table, and the corpuscles are still tied to each other by ether vortices. The continuation of the earlier ideal is confirmed by Thomson's specific reference to Prout's hypothesis not only in this context, but also, prior to the discovery of the electron, in connection with the penetration of matter of different densities by the new x-rays.[15] At this time, in 1896, Rutherford was a research student, under Thomson at the Cavendish Laboratory, working both on radiations and on the conduction of electricity through gases. And although there is only circumstantial evidence of a direct influence of the unifying thread of chemical physics on Rutherford, we at least know of his early interest in 'The Evolution of the Elements'.[16] We also have his word[17] that even before arriving at Cambridge he had read everything that Thomson had written – and that was a great deal.

But as our threads enter the watershed of radioactivity and atomic transmutation, they do not vanish; in a way they are tied to others emerging on the other side, from about 1903 onwards. Firstly, there is a theme which may be called 'cosmical radioactivity': the notion that the sun and stars might owe their high temperatures to an atomic disintegration process. Rutherford and Soddy saw the link with Lockyer's earlier theory and pointed out that theirs was in fact an inversion of his, that atomic disintegration *caused* the observed high temperatures, instead of the reverse.[18]

Secondly, there is a theme that we may designate 'Proutian radioactivity.' The known release of helium from several chemical elements evidently now reinforced Prout's hypothesis, in a modified form. While some, such as Rutherford, were circumspect in suggesting that the helium atom was but '*one* of the secondary units with which the *heavier* atoms are built up',[19] others were less cautious. In 1906 Thomson demolished his own all-electronic atomic theory by showing that 'The Number of Corpuscles in an Atom' was but one thousandth of the number required to account for its mass.[20] At the opposite end of the cathode-ray tube, where he had first extracted his corpuscular substance, Thomson found an alternative. The 'Rays of Positive Electricity', which emerged from a variety of different gaseous chemical elements, were apparently composed of streams of atoms of both hydrogen and helium.[21] But though his dualist Proutian interpretation of these results unfortunately proved short-lived, the general idea projected into the future.

A third theme we can call 'universal radioactivity', the idea that *all* matter, all of the chemical elements, are in the process of radioactive decomposition, albeit very slowly. Evidence for this ranged from apparently reliable direct observations of the activity of ordinary materials;[22] through the 'critical velocity' below which $\alpha$-rays could not be detected; through the strange 'rayless' transmutations detected indirectly;[23] to the oft-mentioned 'general principle' that 'every physical property discovered for one element has been found to be shared by others'.[24] Unviersal radioactivity also fitted in well with the all-electronic atomic theory then in vogue. As Rutherford remarked (1904):

> According to the modern views of the constitution of the atom, it is not so much a matter of surprise that some atoms disintegrate as that the atoms of the elements are so permanent as they apepar to be.[25]

But now it behooves one to ask – what did Frederick Soddy think of all this? He had certainly joined in the common chemical attacks upon Thomson's electron theory in its early days, but now he had become co-founder of the disintegration theory of radioactivity. In 1904, during the course of his dozen lectures on radioactivity, he described quite fully the current cathode-ray and beta-ray researches as well as the theoretical reasoning which led to an electrical explanation of mass or matter, in the mode of the physicists. 'How far these calculations possess a real meaning I am not in a position to say', he remarked.[26] He went a stage further soon afterwards, in his book, with a surprising comment:

> It may at once be pointed out that the theory of atomic disintegration ... is independent of the electrical or electronic view of atomic constitution. It postulates no view of atomic structure beyond the original conception of Dalton.[27]

Having, in his own mind, thus separated the radioactive transformation theory from the theory of electrons, he further attacked the latter with his remark that the theoretical dependence of the electrical mass of an atom upon its internal energy served to show:

> how useless it is to attempt to find numerical relations between the atomic weights of the

elements of the Periodic Table. It is notorious that all such efforts have been fruitless, but it is only recently that the reasons for the failure have been indicated.[28]

Such acidic remarks may be interpreted both as a chemist's claim for independence and as an attempt to confound the physicists with their own labours.

I hope that the present reader would *not* agree with Soddy in this instance, for I believe it *is* of value to bring out such long-term aims and unsolved problems as well as to consider their possible influence.

*Department of Humanities*
*Teesside Polytechnic*
*Middlesbrough Cleveland TS1 3BA*
*England*

## Notes

1 Crookes, W.: 1886, 'Presidential Address to Chemical Section', *Report of British Association for the Advancement of Science*, 558–576, p. 560. He cites several predecessors and his contemporary, Norman Lockyer.

2 *Ibid.*, p. 561.

3 Curie, M.: 1899, 'Les Rayons de Becquerel et le Polonium', *Revue Générale des Sciences* **10**, 41f.; Joliot-Curie, I. (ed.), *Oeuvres de Marie Sklodowska Curie*. Warsaw: Polish Academy of Sciences, 1954, pp. 60–76; my translation.

4 Glazebrook, R. T.: 1885, 'Report on Optical Theories', *Report of British Association*, 157–261, p. 257.

5 Larmor, J.: 1900, *Aether and Matter*, Cambridge, Engl.: University Press, p. vi.

6 Zeeman, P.: 1897, 'On the Influence of Magnetism on the Nature of the Light emitted by a Substance', *Phil. Mag.* **43**, 226–239.

7 Thomson, J. J.: 1897, 'Cathode Rays', *Phil. Mag.* **44**, 293–316.

8 Lodge, O.: 1885, 'On Electrolysis', *Report of British Association*, 723–772; my italics.

9 Thomson, J. J.: 1894, 'The Connection between Chemical Combination and the Discharge of Electricity through Gases', *Report of British Association*, 482–493.

10 Thomson, J. J.: 1883, *A Treatise on the Motion of Vortex Rings*, London: Macmillan & Co. For a fuller discussion see Sinclair, S. B.: 1976, *The Early History of Radioactivity (1896–1904)*, London University Ph.D. Thesis, pp. 32–47.

11 'Strength' = mean velocity of rotation × section area.

12 Thomson, J. J.: 1883, 'On a Theory of the Electric Discharge in Gases', *Phil. Mag.* **15**, 427–434, p. 428.

13 For example, Thomson, J. J.: 1893, *Notes on Recent Researches in Electricity and Magnetism*, Oxford, Engl.: Clarendon Press.

14 See note 6.

15 Thomson, J. J.: 1896, 'The Röntgen Rays', *Nature* **54**, 302–306.

[16] Eve, A. S.: 1939, *Rutherford*, Cambridge, Engl.: University Press, p. 10. A discussion paper in New Zealand, about 1895.
[17] Rayleigh, Baron (Strutt, R. J.): 1942, *The Life of Sir J. J. Thomson*, Cambridge, Engl.: University Press, p. 62. For further discussion of Rutherford's earliest researches see Sinclair, S. B. *op. cit.* (see note 10), pp. 70–91.
[18] Rutherford, E., and Soddy, F.: 1903, 'Radioactive Change', *Phil. Mag.* **5**, 576–591; Chadwick, J. (ed.), *The Collected Papers of Lord Rutherford of Nelson*, London: Allen and Unwin, 1962, Vol. I, p. 608. Anticipated by Stark, J.: 1902, *Die Elektrizität in Gasen*, Leipzig: J. A. Barth, pp. 34–35.
[19] Rutherford, E.: 1905, *Radio-activity*, 2nd Ed., Cambridge, Engl.: Cambridge Natural Science Manuals, Physical Series, p. 484; my italics.
[20] Thomson, J. J.: 1906, 'On the Number of Corpuscles in an Atom', *Phil. Mag.* **11**, 769–781.
[21] Thomson, J. J.: 1907, 'Rays of Positive Electricity', *Royal Institution Library of Science* **6**, 232–247.
[22] For example, Campbell, N. R.: 1905, 'The Radiation from Ordinary Materials', *Phil.* Mag. **9**, 531–544; and *idem*: 1906, *Phil. Mag.* **11**, 202–226.
[23] Summarised in Rutherford. E.: 1905, *Radio-activity*, 2nd Ed., Cambridge, Engl.: Cambridge Natural Science Manuals, Physical Series, pp. 454–455, 549, 552–553.
[24] Schuster, A.: 1903, 'Cosmical Radioactivity', *Report of British Association,* 538–539.
[25] Rutherford, E.: 1904, *Radio-activity*, Cambridge, Engl.: Cambridge Natural Science Manuals, Physical Series, pp. 339–340, Section 270.
[26] Soddy, F.: 1903–1904, 'Radio-Activity', *The Electrician* **52**, 163.
[27] Soddy, F.: 1904, *Radio-Activity: An Elementary Treatise from the Standpoint of the Disintegration Theory*, London; 'The Electrician' Printing and Publishing Co., p. 55.
[28] *Ibid.*, p. 64.

# PART III

# ELEMENTS AND ISOTOPES

NORMAN FEATHER†

# 3. ISOTOPES, ISOMERS, AND THE FUNDAMENTAL LAW OF RADIOACTIVE CHANGE

*When Rutherford and Soddy concluded in 1903 that the radioactive emanations of radium and thorium are 'inert gases of the argon family', they did not ask the pertinent question 'Are these gases to be assigned to the same place in the periodic table, and, if so, what would be the implications of such double occupancy?' This paper summarises the accumulation of evidence resulting ten years later in Soddy's introduction of the term 'isotopes', and it describes Soddy's 1917 prediction of nuclear isomers, the first example of which was discovered by Hahn in 1921. In 1909 Soddy suggested the possibility of alternative disintegration modes, and he correctly predicted their essential features. Within the next ten years, three cases of radioactive series branching were convincingly established. Over many years he gave much thought to the interpretation of the radioactive constant λ – and to the 'cause' of radioactivity. His writings on this subject suggest that he never abandoned his early deterministic philosophy.*

In 1861, at the age of twenty-nine, William Crookes discovered a new element which he named thallium, thereby commemorating the green line in the spectrum which revealed its existence. In the wisdom of old age, for he was then turned seventy, Michael Faraday wrote in congratulation 'To discover a new element is a very fine thing, but if you could decompose an element and tell us what it is made of – that would be a discovery indeed worth making'. Just about forty years later, in 1903, to be precise, Ernest Rutherford, the physicist, and Frederick Soddy, the chemist, took the first step towards that greater discovery – the step of recognition: they recognised that the atoms of the radioactive elements are all the time decomposing themselves. [1]

By that time Rutherford and Soddy had indeed discovered a new element – or, perhaps, more than one. Some ten years previously John William Strutt, the third Lord Rayleigh, and William Ramsay had discovered argon; [2] effectively, by 1898, a whole new column had been added to the periodic table of the chemists – it was required to accommodate the five gaseous elements helium, neon, argon, krypton, and xenon, which Ramsay and his collaborators had by then identified in the

*George B. Kauffman (ed.), Frederick Soddy (1877–1956)*, 57–65.

earth's atmosphere. In April 1902 Rutherford and Soddy had written, regarding their experiments with the thorium emanation, '. . . the interpretation of the experiments must be that the emanation is a chemically inert gas analogous in nature to members of the argon family'. [3] The emanation was their new element. In the following year they studied even more extensively the emanation from radium compounds, [1] the existence of which Dorn had been the first to announce. [4] They reported 'The experiments already described on the chemical nature of the thorium emanation were repeated for that of radium. As in the former case, all the reagents tried were without effect'. They observed some difference in behaviour of the two emanations in relation to condensation and volatilisation, but they were so convinced that they were in fact dealing with two substances of closely identical chemical properties that they wrote 'This difference . . . is explained on the view that the number of particles of emanation present for equal effects is probably many thousand times greater in the case of radium emanation than in the case of thorium emanation'. A year later Soddy wrote, unreservedly, 'The conclusion is indicated that the [radium] emanation is a chemically inert element of the argon family'.

I have not recited the history of this episode in order to highlight the unfailing intuition of these young experimenters – for we now know that, in this context, all their conclusions were well-founded. My purpose is, rather, to draw attention to a simple fact. Rutherford and Soddy had correctly identified two substances of different origins, and of markedly different radioactive properties, each as a higher homologue of the inert gases of column 0 of the periodic table. On the other hand, there is not the slightest evidence that either of them asked himself the question 'Can there possibly be two vacant places in column 0 available for these elements – and, if not, what are the implications of a doubly-occupied place?' The cogency of this hypothetical question was intensified once it was accepted that the emanation from actinium compounds, independently discovered by Giesel [5] and Debierne, [6] had added a third, radioactively-distinct, inert gas to this group. However, the question was not asked. In truth, the time was not ripe. There was no sufficient basis of accepted theory, and there were so many other problems to engage their attention.

Six or seven years later, Soddy was beginning to ask himself that question, but I must leave it to Professor Badash (Paper 1) and Dr. van Spronsen (Paper 5) to lead you to his answer and to the concept of isotopes announced, almost casually, in a letter to *Nature* in December 1913. [7]

In a Royal Institution discourse on May 18, 1917 Soddy speculated further on the concept of isotopy. In the same vein as in his *Nature* letter of four years previously, he wrote 'The mass depends on the *gross* number of + charges in the nucleus, chemical properties on the difference between the gross numbers of + and − charges. But the radioactive properties depend not only on the gross number of charges, but on the constitution of the nucleus. We can have isotopes with identity of atomic weight as well as of chemical character, which are different in their stability and mode of breaking up'. This last statement, when it was written, was pure speculation.

Soddy returned to the subject in a lecture to the Chemical Society on December 19, 1918. He said, 'Isotopes need not, however, have different atomic weights.... Isobaric isotopes of the character in question can only at present be distinguished if they are unstable.... The accomplishment of artificial transmutation would reveal them if they existed ... and the discovery of any new property ... concerned with the nucleus of the atom rather than its external shell, might also be a means of revealing differences of this character'. This was still in the realms of speculation – indeed Soddy was here speaking of artificial transmutation (without saying how it might be achieved) six months before the publication of Rutherford's paper [8] on 'An anomalous effect in nitrogen'. However, within three years Hahn had discovered uranium Z, [9] and his own further investigations [10] and those of Guy [11] and Russell, [12] in 1923, made it abundantly clear that this body, radioactively quite distinct from uranium $X_2$, was both isotopic and isobaric with it. Soddy's speculation had been vindicated – even though this single example remained unique for another sixteen years. Then, as a result of the study of artificial transmutations, further examples of similar pairings began to be discovered and to multiply, until today they are numbered in hundreds. Soddy had originally used the terms isobaric and heterobaric to designate pairs of nuclei of the same and of different mass numbers, respectively. Nowadays, we use the term isomers, borrowed from the organic chemists, when referring to nuclei of which the mass and charge numbers are both the same. I am not sure who introduced the term in its nuclear context; I am inclined to believe that it was adopted without formal commitment. Its first appearance in print – in its adjectival rather than nominal form – as far as I am aware, is in the published version of a paper which George Gamow contributed to an international conference in Cambridge in October, 1934.

I now pass to the main topic of my paper – the problem of the fundamental law of radioactive change. 'Radioactive change', as is well known, was the simple title of the now classic paper of May, 1903 [1] in which Rutherford and Soddy gave final shape to the disintegration hypothesis which had been forming in their minds during the preceding year. They wrote:

> ... the law of radioactive change ... may be expressed in the one statement – the proportional amount of radioactive matter that changes in unit time is a constant.... [this] constant $\lambda$ ... possesses for each type of active matter a fixed and characteristic value. $\lambda$ may therefore be suitably called the 'radioactive constant'.... The law of radioactive change, that the rate of change is proportional to the quantity of changing substance, is also the law of monomolecular chemical reaction. Radioactive change, therefore, must be of such a kind as to involve one system only,... the changing system must be the chemical atom ... in radioactive change the chemical atom must suffer disintegration.

Nothing in this statement is questioned today.

Soddy gave much thought, over the years, to the implications of this simple law. On February 23, 1904 he gave the Wilde Lecture before the Literary and Philosophical Society in Manchester. [13] Discussing the law then, he said 'A relation perhaps more generally useful than the radio-active constant is its reciprocal $1/\lambda$ which has a very interesting physical significance. It represents the average life of the metabalon....' [13] 'Metabalon' was a term introduced by Rutherford and Soddy to distinguish a radioactive element from a stable one. It did not find favour with scientists generally, and its use was soon abandoned. Soddy continued, 'The average life of a metabalon may be compared with the atomic weight in the case of a stable atom as a constant well suited for its experimental identification'. [13] This was a percipient remark: it carried the germ of the prediction made thirteen years later, and which we have already discussed, that atomic species may exist having different radioactive constants – that is, different mean lifetimes – but the same atomic weight.

At this stage in his lecture Soddy made a remark which was to characterise his whole attitude to the problem in the future. It consisted of a statement of fact, and a comment on that statement. The fact was incontrovertible; the comment disclosed the philosophy of the speaker. It was the philosophy of determinism, which in the years ahead was to become less and less acceptable to the physicists of the day. Soddy said,

'It may be pointed out that the actual life of the different atoms of the same unstable element has all values between zero and infinity' – [13] that was the true statement; then the comment, 'This constitutes the first difference in properties between the individual atoms of the same element that has ever been discovered'. [13] Note the term 'properties'. Soddy had an unrivalled command of the English language; when he wrote 'properties', we must believe that the word matched his intentions as precisely as the language permitted.

Do not imagine that I am belittling Soddy's genius as experimental scientist by exposing his philosophy, in the year 1904, as deriving essentially from the nineteenth century. He was not alone in this attitude. At the Solvay Conference in Brussels nine years later, Marie Curie said (I translate from the original French):

> And since, in a sample containing a large number of atoms, some break up almost at once, while others can persist for a very long time, it is no longer possible to regard all the atoms of the same simple substance as similar in all respects, rather one must accept that differences in their future history are determined by individual differences [already existing between them].

Again, it must be admitted that Soddy was not unaware of a possible alternative point of view, even in 1904. In his Manchester lecture he had written 'Radioactivity is a property contributed by a few atoms only in any given instant. . . . For many purposes, however, a property which is contributed by a *constant* fraction of the total is indistinguishable from a property possessed by each atom in common'. [13] Here he pointed the way to the alternative viewpoint, but he did not pursue it: it had no place in his philosophy.

It was otherwise in Vienna. Von Schweidler, a collaborator of Stefan Meyer for many years, posed the significant question: Given the accepted doctrine that the instant of break-up of any particular atom is unpredicable, $1/\lambda$ being merely the average survival lifetime for a very large number of similar atoms, can it be possible that, in respect of a radioactive source containing a finite number of atoms, the fraction of those atoms breaking up in each succeeding unit interval of time is strictly constant and of magnitude $\lambda$? He was convinced that the answer was 'No'. There must be fluctuations in the value of this fraction, relatively the more pronounced as the disintegration-rate is smaller.

In 1905 von Schweidler published a short paper giving the theory of those fluctuations. He did not argue the point: he simply introduced his calculation with the bold statement, 'If $\lambda\, dt$ is the probability that an atom suffers disintegration within the interval $dt$, then the probability that it survives unchanged throughout time $t$ is equal to $e^{-\lambda t}$ and $\tau = 1/\lambda$ is its average lifespan' [14] (I translate from the German). In essence, he had transferred the radioactive property measured by $\lambda$ from the many atoms of the source, viewed collectively, to each individual atom. For him, $\lambda$ had become an atomic constant, rather than the measure of a property of a particular type of matter in bulk. On the basis of this new interpretation, $\lambda$ was also the most probable, not the uniformly constant, value of the fractional disintegration rate for any particular sample of the radioelement concerned.

Von Schweidler was ahead of his time in two respects. In the first place, it was some years before experimental technique had developed to the stage when his theoretical predictions could be adequately tested (when the experiments were eventually performed, the predictions were cofirmed in their entirety); in the second place, the fundamental change of emphasis in the interpretation of $\lambda$ was of such novelty that its true character did not strike home for another quarter of a century or thereabouts – to his readers, and possibly to von Schweidler himself, it had the appearance of an interesting mathematical transformation and nothing else. Physicists could still ask the question 'What is the cause of radioactivity?', and hope for an intelligible answer – though some, it must be admitted, and Rutherford was a prime example, were content to let the question lie, and get on with their experiments.

Some years later Soddy was still giving thought to the problem. In a paper published in November, 1909 he wrote:

> The cause of atomic disintegration remains unknown. It is difficult to construct any model of the disintegrating mechanism.... What may be termed the inevitableness of the process, and its entire independence of all known conditions, suggests that the cause of disintegration is apart from the atom. It is difficult to believe that the cause is resident in space external to the atom. It seems more probable that it exists within the atom and at the same time is uninfluenced by it.

This phraseology makes no obvious sense to us today. It seems to me to convey an echo, in a sub-atomic rather than a molecular context, of the chemist's picture of molecular activation, when the reaction is monomolecular. It must not be forgotten that Rutherford and Soddy used the analogy of the monomolecular chemical reaction when presenting the disintegration hypothesis to the world. Soddy was well aware of the limitations of this analogy – in 1904 he had pointed out that monomolecular chemical reactions are nearly all endothermic, whilst radioactive disintegration is necessarily exothermic; in the former case, energy has to be acquired from the surroundings; the molecules have to be activated before they can contribute to the process; in the latter case, the energy must be there already, though it may need a trigger action, or catalyst, to release it. In the molecular case, the thermal chaos of the molecules of the surrounding medium is the source of the random supply of activation energy. It seems to me that Soddy was postulating an aethereal chaos within the radioactive atom which would provide the random trigger action necessary to initiate its disintegration.

I now skip a period of almost twenty years, and remind you that one of the uncovenanted successes of the then newly developing wave mechanics was the formulation of the first successful description of radioactive $\alpha$-particle emission in terms of a simple mathematical model. Gurney and Condon in America, and Gamow in Europe, independently made the break-through. It was obvious at once that, the value of the energy of $\alpha$-disintegration being given, a value of the disintegration constant $\lambda$ could be calculated in terms of the charge on the nucleus and the nuclear radius, the value of Planck's constant being assumed. Even on the basis of the simplest model, the measure of success was altogether convincing. It was clear that, in principle, if more were known concerning the forces effective within the nucleus, the disintegration energy itself could be calculated, and thence $\lambda$ – indeed the whole process could be derived, in relation to its measurable characteristics – 'from first principles', as they say. For the purposes of our present discussion, there is just one point to note – wave-mechanical theory enables the radioactive constant $\lambda$, for $\alpha$-disintegration, to be calculated in terms of quantities characterising the individual atomic nucleus. $\lambda$ is indeed, intrinsically, a probability constant having direct relevance to the individual atom. The many atoms in a pure radioactive source have different lifetimes, but they have the same $\lambda$ – in precisely the same sense, formally, as they have the same mass, or contain the same number of electrons. This, of course, is the von Schweidler view, [14] up-dated with the times.

By 1929 when the wave-mechanical calculation was being accorded universal acceptance by workers active in the field, Soddy's original contributions to research in radioactivity were coming to a close. His last contribution was a letter to *Nature* in 1932. [15] It would be possible for me to end my story at this point, leaving unexplored the question of his reaction, if any, to the newer ideas during the remaining years of his long life. There is, however, a postscript, and, because I was innocently responsible for its being written, I feel under some obligation to record it here.

In the autumn of 1939 I was revising the first draft of my life of Lord Rutherford, [16] and to ease my conscience I wrote to all those whose work I had quoted – at least, to those surviving – to make sure that in my associated comment I had not misrepresented their views. I wrote to Soddy, sending him a copy of my draft account of the early development of opinion regarding the significance of the radioactive constant $\lambda$. A correspondence of some three weeks' duration ensued. In the upshot I modified my draft in certain particulars, though I did not achieve complete agreement with my correspondent on all the issues involved. The postscript I referred to is provided by his last letter to me, dated October 3, 1939. In it he wrote:

> But I have still to learn that there has been any real advance in 'explanation' of the view I reached and put forward in 1904. We part company, I think, at von Schweidler's 'Let the probability of an atom breaking up in the unit time be *p*', thus endowing 'an ignorance of the whole matter' as Poincaré would say with a simulation of mathematical precision, as though *p* referred to pennies or pigs instead of what I termed chance, though I did not call it *c*, I admit.... However, thank God I'm no mathematician!

Soddy, indeed, had the last word.

*Department of Physics*
*University of Edinburgh*
*Maxwell Building*
*Edinburgh EM9 3J2*
*Scotland*

## Bibliography

[1] Rutherford, E., and Soddy, Frederick: 1903, 'Radioactive Change', *The London, Edinburgh, and Dublin Philosophical Magazine and Journal of Science* [6] **5**, 576–591.

[2] Preliminary announcement, British Association for the Advancement of Science, Oxford, August 13, 1894. Rayleigh, Lord, and Ramsay, W.: Feb. 1, 1985, 'Argon: A New Constituent of the Atmosphere', *Chem. News* **71**, 51–58.

[3] Rutherford, E., and Soddy, Frederick: 1902, 'The Radioactivity of Thorium Compounds. I. An Investigation of the Radioactive Emanation', *J. Chem. Soc.* **81**, 321–350.

[4] Dorn, F. E.: 1900, 'Von radioactiven Substanzen ausgesandte Emanation', *Abh. Naturf. Ges., Halle* **23**, 1, 1–15.

[5] Giesel, F. O.: 1902, 'Ueber Radium und radioactive Stoffe', *Ber.* **35**, 3608–3611; 'Ueber den Emanationskörper aus Pechblende und über Radium', *ibid.*, 1903, **36**, 342–347; 'Ueber den Emanationskörper (Emanium)', *ibid.*, 1904, **37**, 1696–1699, 3963–3966; 1905, **38**, 775–778; 1907, **40**, 3011–3015.

[6] Debierne, A.: Oct. 16, 1899, 'Sur une nouvelle matière radioactive', *Compt. rend.* **129**, 593–595; 'Sur une nouvel élément radioactif: l'actinium', *ibid.*, Apr. 2, 1900, **130**, 906–908; 'Sur du baryum radioactif artificiel', *ibid.*, July 30, 1900, **131**, 333–335; Feb. 16, 1903, **136**, 446–449; Mar. 16, 1903, **136**, 671–673; 'Sur l'émanation de l'actinium', *ibid.*, Feb. 15, 1904, **138**, 411–414; 'Sur l'actinium', *ibid.*, Oct. 3, 1904, **139**, 538–540; 'Sur les gas produits par l'actinium', *ibid.*, Aug. 14, 1905, **141**, 383–385.

[7] Soddy, Frederick: Dec. 4, 1913, 'Intra-atomic Charge', *Nature* **92**, 399–400.

[8] Rutherford, E. E.: 1919, 'Collision of Alpha Particles with Light Atoms. III. An Anomalous Effect in Nitrogen', *Phil. Mag.* **37**, 581–587.

[9] Hahn, O.: June 11, 1921, 'Über eine neue radioaktive Substanz im Uran', *Ber.* **54**, 1131–1142.

[10] Hahn, O.: 1923, 'Über das Uran Z und seine Muttersubstanz', *Z. physik Chem.* **103**, 5, 6: 461–480.

[11] Guy, W. G., and Russell, A. S.: 1923, 'The Short-Lived Radioactive Products of Uranium', *J. Chem. Soc.* **123**, 2618–2631; *Nature* **111**, 703.

[12] Russell, A. S.: 1923, 'The Relation of Actinium to Uranium', *Nature* **111**, 703–704.

[13] Soddy, F.: 1904, 'The Evolution of Matter as Revealed by the Radio-active Elements', *Manchester, Mem. Lit. Phil. Soc.* **48**, 1–42.

[14] Schweidler, E. von: 'Über Schwankungen der radioaktiven Umwandlung', *Compt. rend. Prem. Congr. Rad. Ion. (1905)*, Brussels: 1906, pp. 1–3, p. 1; reprinted in Trenn, Thaddeus J.: *The Self-Splitting Atom: The History of the Rutherford-Soddy Collaboration*, London: Taylor & Francis, 1977, pp. 157–159.

[15] Soddy, Frederick: 1932, 'The $\alpha$-Rays of Ionium', *Nature* **130**, 364–365.

[16] Feather, Norman: 1940, *Lord Rutherford*, London: Blackie & Son.

GEORGE B. KAUFFMAN

# 4. THE ATOMIC WEIGHT OF LEAD OF RADIOACTIVE ORIGIN: A CONFIRMATION OF THE CONCEPT OF ISOTOPY AND THE GROUP DISPLACEMENT LAWS*

*The development of the concept of isotopy from the discovery of radiothorium by Otto Hahn in 1905 to Soddy's proposal of the term 'isotope' in 1913 is presented along with a brief review of the Fajans-Soddy group displacement laws. This is followed by a discussion of the work of Kasimir Fajans, Frederick Soddy, and others in verifying two major predictions of these laws, viz., the origin of actinium and the atomic weight of lead of radioactive origin. Within about a single month in 1914 the existence of isotopes by determination of the atomic weight of lead of radioactive origin was confirmed experimentally in four independent laboratories in four different countries – Soddy and Hyman (England, May 7), Richards and Lembert (USA, May 21), Curie (France, June 8), and Hönigschmid and Horovitz (Austria, June 15) – a tribute to the international character of science. Richards and Lembert's work has been called 'the final act of the dramatic events surrounding the clarification of the place of the radioelements in the periodic table'.*

## 1. The Concept of Isotopy

The concept of isotopy is inextricably linked with the name of Frederick Soddy (1877–1950),[1–5] who first proposed the term 'isotope' in 1913. Nevertheless, the idea of different species of the same element evolved during the decade preceding Soddy's definitive statement of the concept.

In 1902 Ernest Rutherford and Soddy[6] proposed their disintegration theory of radioactivity in which they attributed the phenomenon to a spontaneous disintegration of the radioactive element by expulsion of particles, resulting in formation of new elements – a modern variation on the ancient alchemists' dreams of transmutation. In July, 1903 William Ramsay and Soddy, working in London, experimentally confirmed the theory's prediction that radium should continuously produce helium.[7] According to Soddy himself, 'the history of isotopes fittingly commences with the 1905 discovery of radiothorium,'[8] a product in the

*George B. Kauffman (ed.), Frederick Soddy (1877–1956)*, 67–92.

thorium disintegration series, by Otto Hahn, working under Ramsay's direction.[9] Attempts to separate the new substance from thorium compounds were either unsuccessful or successful to only a small extent. For example, in 1906 Julius Elster and Hans F. K. Geitel[10] of Wolfenbüttel and G. A. Blanc[11] of Rome, who had discovered radiothorium in radioactive sediments from hot springs, apparently succeeded in their attempts to only a small extent. Bertram Borden Boltwood,[12] originally a consulting chemist at New Haven, Connecticut and then a professor at Yale University, H. M. Dadourian[13] also of Yale and Herbert Newby McCoy and W. H. Ross[14] of the University of Chicago also made unsuccessful attempts at separation. In 1907 Hahn[15] discovered a disintegration product intermediate between thorium and radiothorium, which produces radiothorium. In the following year he[16] showed that this product, which he had named mesothorium, consists of two successive products – mesothorium 1 and mesothorium 2.

In a paper[17] on the specific radioactivity of thorium dated October 12, 1907 containing data obtained during the years 1905 through 1907, McCoy[18,19] and Ross converted eight commercial samples of thorium nitrate to thorium dioxide (thoria) by precipitation with various reagents (the number in parentheses indicates the number of precipitations) – oxalic acid (1 and 8), sodium thiosulfate (8), potassium chromate (8), barium oxalate (7), ammonia (1 and 100), hydrogen peroxide (40) – followed by ignition of the precipitates. After measuring the decline in activities of the resulting thoria samples with time, they stated that 'our experiments strongly indicate that radiothorium is entirely inseparable from thorium by chemical processes'.[20] Soddy considered this 'the first definite statement of the doctrine of the complete chemical non-separability of what are now called isotopes',[21] in other words, 'the complete chemical identity of isotopes, as distinct from close chemical similarity'.[22] McCoy and Ross correctly concluded from their results that 'the isolation of radiothorium from thorianite and from pure thorium nitrate . . . may have been accomplished by the separation of mesothorium, which in time changed spontaneously into radiothorium'.[23]

Evidence for pairs or groups of chemically nonseparable radioactive elements now began to accumulate. In 1907 Soddy wrote, 'Although the separation of mesothorium from thorium and of thorium X from radiothorium is easily accomplished, there is no known method of separating by chemical means either radiothorium from thorium or thorium X from mesothorium'.[24] In 1908 Boltwood[25] and, independently, Willy Marckwald and Bruno Keetman[25] at the University of Berlin discovered ionium, which was found to be chemically identical with thorium. For example, in 1909 Keetman[27] reported his failure to separate ionium from thorium by any of twelve methods known to be effective for purifying thorium. In 1910 Baron Auer von Welsbach[28] reported a successful separation of ionium from actinium, but he failed to separate ionium from thorium. With respect to Boltwood's, Keetman's, and Auer von Welsbach's unsuccessful attempts at separation, Soddy wrote:

> The question naturally arises whether some of the common elements may not, in reality, be mixtures of chemically non-separable elements in constant proportions, differing stepwise by whole units in atomic weight. This would certainly account for the lack of regular relationships between the numerical values of the atomic weights.[29]

In 1910 H. Herchfinkel[30] at Madame Curie's laboratory unsuccessfully attempted to concentrate radium D (radio-lead) from ordinary lead, and in 1913 F. A. Paneth and G. von Hevesy[31] likewise failed to accomplish this even though they used twenty different methods. From their failure they devised the extremely versatile and elegant method of employing radioactive isotopes as tracers.[32]

In 1909 Daniel Strömholm and The Svedberg at the Uppsala University employed Eilhard Mitscherlich's 1819 law of isomorphism to characterize chemically some of the radioactive elements by crystallizing different salts in solutions of salts of these elements and observing whether or not these elements crystallized along with the salts. For example, inasmuch as thorium X crystallizes along with barium and lead salts, but not with potassium, bismuth, and lanthanum salts, they concluded that it is an alkaline earth element (Group II of the periodic table).[33] They likewise characterized as an alkaline earth element actinium X, a radioactive disintegration product intermediate between actinium and its emanation, i.e., the actinium series analogue of thorium X, which had been discovered in 1905 by Tadeusz Godlewski of the Technische Hochschule, Lemberg.[34] By the same method they could not find any chemical differences between actinium X, thorium X, and radium.

According to Soddy, 'Strömholm and Svedberg were probably the first to attempt to fit a part of the disintegration series into the Periodic Table' and 'in their conclusion, is to be found the first published statement, that the chemical non-separability found for the radio-elements may apply also to the non-radioactive elements in the Periodic Table'.[35] They were also 'the first to suggest a general complexity of the chemical elements concealed under their chemical identity'.[36] Recognizing that there are three independent but parallel series of radioactive elements (uranium, thorium, and actinium series), Strömholm and Svedberg concluded,

> one may suppose that the genetic series proceed down through the periodic table, but that always the three elements of the different genetic series, which thus together occupy one place in the periodic system, are so alike that they always occur together in nature and also have not been able to be separated appreciably in the laboratory. Perhaps, one can see, as an indication in this direction, the fact that the Mendeleev scheme is only an approximate rule as concerns the atomic weight, but does not possess the exactitude of a natural law; this woud not be surprising if the elements of the scheme were mixtures of several homogeneous elements of similar but not completely identical atomic weights.[37]

In suggesting that the presence of what were later called isotopes might not be limited to the radioactive elements, Strömholm and Svedberg offered a possible explanation for the deviation of atomic weights from integral values and raised the point that Prout's hypothesis might possess some validity. Of course, it was not until 1913 that Henry Gwyn Jeffries Moseley[38] showed that atomic numbers rather than atomic weights should be the basis for placing elements in the periodic table.

In 1910, in a study of a radium preparation prepared from a uranium mineral containing thorium, Marckwald[39] demonstrated the chemical inseparability of mesothorium 1 and radium, a conclusion confirmed by Soddy's investigation of preparations obtained from thorianite.[40] In the same study Soddy also confirmed Strömholm and Svedberg's nonseparability of thorium X, mesothorium, and radium. In his review of their work, Soddy concluded that such nonseparability of the radioactive elements was based not on close similarities but rather on complete chemical identities:

> When it is considered what a powerful means radioactive methods of measurement afford for detecting the least change in concentration of a pair of active substances, and the completeness and persistence of some of the attempts at separation, which have been made, the conclusion is scarcely to be resisted that we have in these examples no more chemical analogues but chemical identities.[41]

This was an important assertion, for nonidentical but chemically nonseparable elements were then known, *e.g.*, the rare earths. In Hahn's words, 'Soddy possessed the courage to declare that these were chemically identical elements'.[42] In his 1910 annual report Soddy, in connection with Strömholm and Svedberg's work, also stated:

The recognition that elements of different atomic weight may possess identical chemical properties seems destined to have its most important application in the region of inactive elements.... Chemical homogeneity is no longer a guarantee that any supposed element is not a mixture of several of different atomic weights, or that any atomic weight is not merely a mean number. The constancy of atomic weight, whatever the source of the material, is not a complete proof of homogeneity, for, as in the radio-elements, genetic relationships might have resulted in an initial constancy of proportion between the several individuals, which no subsequent natural or artificial chemical process would be able to disturb.[43]

In 1912 Sir J. J. Thomson applied his positive ray method to atmospheric neon of known atomic weight 20.2 and discovered two curves – the expected one corresponding to atomic mass 20 and an unexpected one, much fainter than the first, corresponding to atomic mass 22.[44] Being hesitant to admit that he had discovered two forms of neon with different atomic weights, Thomson stated that neon was a mixture of two gases but that the heavier one, which he called metaneon, might be an unknown compound of the single neon species. Thomson's assistant, Francis W. Aston, of later mass spectrograph fame, employed fractional gaseous diffusion to show that atmospheric neon is not homogeneous but can be separated partially into fractions of different density, which, however, exhibit identical spectra and appear to be identical in all properties other than atomic weight.[45] Soddy did not share Thomson's hesitation but declared that Thomson's and Aston's work on neon

appears to be a case of isotopic elements outside the radioactive sequences.... The discovery is a most dramatic extension of what has been found for the elements at one extreme of the periodic table, to the case of an element at the other extreme, and strengthens the view that the complexity of matter in general is greater than the periodic law reveals. Although the complexity is greater, the problem of atomic structure has been much simplified, because the generalization gives a probable explanation of the absence of exact simple numerical relations among the atomic weights.[46]

The number of known chemically nonseparable elements continued to increase.[47] In a short letter to the Editor of *Nature*, published on December 4, 1913, Soddy first used the term isotope:

The same algebraic sum of the positive and negative charges in the nucleus, when the arithmetical sum is different, gives what I call 'isotopes' or 'isotopic elements,' because they occupy the same place in the periodic table.[48] They are chemically identical, and save only as regards the relatively few physical properties which depend upon atomic mass directly, physically identical also.[49]

According to Soddy's recollection, he coined the term when he 'got tired of writing "elements chemically identical and non-separable by chemical methods"'.[50] According to Soddy's former student Lord Alexander Fleck,

We owe the word 'Isotopes' to Dr. Margaret Todd, a medically qualified Glasgow friend of Mr. and Mrs. Soddy who made her name as a writer using the pen name of Graham Travers. This word was coined at a dinner party – I have always understood a normal, informal Sunday evening meal in the home of Sir George Beilby.[51]

According to Thadeus J. Trenn, perhaps the first use of the term isotope for the position of elements dates from 1893.[52] The Polish chemist Kasimir Fajans (1887–1975), formerly Professor emeritus at the University of Michigan, Ann Arbor, proposed the term 'pleiad', after the conspicuous group of stars in the constellation Taurus, in March, 1913, more than half a year before Dr. Todd suggested the term 'isotope'.[53,54] Thus, although Soddy was neither the first to use the term 'isotope' nor even the first to suggest a specific name for elements with the same atomic number but with different atomic weights, his consistent use of the term has 'ensured its passage into the language of science'.[55]

## 2. The Group Displacement Laws

As stated above, in 1902 Rutherford and Soddy, working at McGill University, Montreal, Canada, proposed in their theory of radioactive disintegration that radioactive elements are transformed into each other and exhibit genetic relationships or decay series. Since the French physical chemist Jean Baptiste Perrin had suggested in the previous year that the change might be occurring in the atoms themselves, Rutherford and Soddy hypothesized that very short-lived, radioactive species are sometimes elements that are transformed into each other by radioactive decay and emission of material particles. Their disintegration theory was so complete that in 1909 only an extension to branching series was required.[56,57] During the ensuing years Rutherford and co-workers at Manchester University, Soddy and co-workers at Glasgow University, and others, including Boltwood, Bronson, Fajans, Fleck, Geiger, Giesel, Hahn, von Hevesy, von Lerch, Marsden, Meitner, Meyer,

Paneth, Ramsay, Russell, and von Schweidler, worked at elucidating the genetic relationships among the radioactive elements, observing their chemical properties, and identifying their modes of disintegration.[58] In the course of these developments, as we have seen, it was observed that some of the newly discovered radioactive substances possessed similar, if not identical, chemical properties and were chemically nonseparable, *i.e.*, they were isotopes.[59] At the same time, as more and more was learned about relationships in the three radioactive decay series (uranium, thorium, and actinium), generalizations attempting to correlate the type of radioactive decay and position in the periodic table of the parent element with the chemical properties of the daughter element product were proposed by von Lerch,[60] Lucas,[61] Cameron,[62] Soddy,[63] and Russell.[64] Such rules, however, were not universally applicable, as was recognized by Fajans[65] and von Hevesy.[66]

The historical development of the so-called Fajans-Soddy group displacement laws has been discussed in a masterly fashion by Lawrence Badash,[67] who, while crediting Soddy with the initial formulation of the alpha decay rule,[68] ascribes the final, correct interpretation of the group displacement laws to Fajans alone.[69,70] Rather than recapitulate Prof. Badash's developments here, we need only concern ourselves with the facts directly related to our theme.

In his *Habilitationsschrift* (Technische Hochschule, Karlsruhe) dated December, 1912 Fajans observed that

> In all cases of beta disintegration the resulting product is indeed electrochemically more negative than its parent. Upon alpha emission, however, the direct opposite is true.[70]

On the last day of the year 1912 Fajans submitted two articles on his displacement laws to the *Physikalische Zeitschrift*, which published them on February 15, 1913:[71]

> (1) Alpha particle emission is accompanied by a transition from right to left in a horizontal row of the periodic table. The observation by Soddy that this proceeds by a jump in the next but one group is assumed to be true in every case.
> (2) In similar fashion it is derived for beta disintegrations that they cause a transition to the next higher group, *i.e.*, from left to right in a horizontal row.[72]

In the first paper Fajans applied his two laws to all those cases in which they could be tested electrochemically, while in the second paper he arranged in the periodic table all the radioactive elements then known and predicted atomic weights for each of them. He accomplished this by using his displacement laws plus the facts, discovered by Ramsay and Soddy[73] and by Rutherford,[74] that alpha particles are dipositive helium ions of atomic weight 4 and that the mass of a beta particle is negligible. In modern terms,

$$ {}^{\text{At. Wt.} - 4}_{\text{At. No.} - 2}\mathrm{Y} \xleftarrow{\alpha} {}^{\text{At. Wt.}}_{\text{At. No.}}\mathrm{X} \xrightarrow{\beta} {}^{\text{At. Wt.}}_{\text{At. No.} + 1}\mathrm{Z} $$

In devising this arrangement, Fajans observed that some radioactive elements of different origin, different properties, and different atomic weights must occupy the same place in the periodic table, forming what he was later to call a 'pleiad'[75] and what Soddy was to call still later 'isotopic elements'.[76] Fajans, Soddy, and others now began work to verify two major predictions of the displacement laws – the origin of actinium and the atomic weight of lead of radioactive origin.

## 3. The Origin of Actinium

Uranium I decays by alpha emission to uranium X, which Fajans assigned to the place occupied by thorium. Uranium X then decays by beta emission to uranium II. The displacement laws predicted that the latter transformation should occur by stepwise emission of two beta particles, with intermediate formation of a radioactive element which should belong in Group V ('eka-tantalum'):

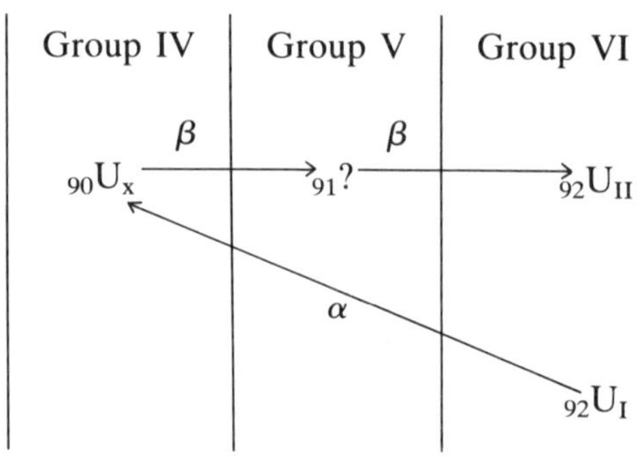


The existence of this new member of the uranium series, which should furnish a key to the origin of actinium, was predicted independently by Fajans, Russell, and Soddy. Reasoning that this element, in analogy with tantalum, should be separable electrochemically from uranium X, Fajans, together with his student Oswald Helmut Göhring, isolated the first known isotope of the new element, which they called uranium $X_2$, by plating it out from uranium X solutions on freshly cleaned lead dishes.[77] Because of its short half-life (1.15 min.), they called the new element 'brevium'. Their discovery proved the validity of the displacement laws and the electrochemical generalizations. It was particularly important in that the new element belonged to a previously unassigned place in the periodic table (atomic number 91), whereas, previously discovered isotopes had all belonged in places already occupied by known elements.

Soddy's work on the origin of actinium proceeded on somewhat different lines. If actinium (at. no. 89) resulted from a beta particle emission, its parent would have to be an isotope of radium (at. no. 88). Soddy disproved this possibility by examining a sample of Friedrich O. Giesel's radium bromide kept for ten years which showed a total absence of radioactinium.[78] The alternative, that actinium is produced by an alpha ray emission from 'eka-tantalum', was finally established in 1918 independently by Soddy and John A. Cranston[79] and by Otto Hahn and Lise Meitner.[80] Soddy and Cranston used uranium $X_2$, which could be volatilized from uranium $X_1$, as a radioactive indicator of the chemical properties of the unknown element. From pitchblende they obtained a sublimate initially free from all the known pre-emanation members but continuously generating actinium with the passage of time. Hahn and Meitner, whose work overshadowed that of Fajans and Göhring, discovered the new element in insoluble residues produced by the separation of radium from pitchblende. They isolated it by methods used to remove tantalum, which was added in a minute amount to facilitate the separation. They proposed the name protactinium, the name by which element 91 is now known.

## 4. The Atomic Weight of Lead of Radioactive Origin: Determinations During 1914

The discovery of protactinium confirmed the group displacement laws, but, a more important confirmation of both the concept of isotopy and of the group displacement laws was sought in atomic weight determinations of isotopic elements. Unfortunately, hardly any radioactive element was obtainable in isotopic purity or quantity sufficient for such a determination. However, the lead end-product of the uranium disintegration series and the lead postulated as the end-product of the thorium series were felt to be ideal for atomic weight determinations.[81,82] In Fajans' words:

A discrepancy still remains to be explained. It is known that the experimentally determined atomic weight of lead (207.1) does not agree with that calculated (206.5) on the basis of the assumption that lead represents the end product of the uranium series.... $ThD_2$, which has a higher atomic weight (208.4) than the lead constituted from uranium (206.5), must be very thoroughly like it.[83] If ordinary lead represents a mixture of these two not easily separable kinds of lead, the discrepancy referred to would be explained.... To test this assumption, atomic weight determinations should be carried out on lead ... which [has] been extracted from thorium-free uranium minerals and from uranium-free thorium minerals.[84]

Elsewhere Fajans wrote:

There will be an apparent difference between the two leads: the atomic weight of one differs from that of the other by two full units.[85]

Soddy made similar predictions:

Both end-products [of the thorium series] ... should be non-separable from lead, the calculated atomic weight of this 'lead' being about 208.5 in each case.... If we suppose that all the lead in the world is produced as the end-products in the three disintegration series in constant proportion, the found atomic weight, 207.1, indicates that about half of it may result in that of thorium and the other half in that of uranium. It is, however, hardly profitable to go further in detail until the constancy of the atomic weight of lead from a variety of radio-active minerals has been experimentally tested.[86]

Soddy questioned whether the supposed end-products of the thorium series would 'be sufficiently stable to accumulate in minerals and to constitute a perceptible fraction of the lead occurring therein'.[87] He therefore warned that

although the atomic weight of lead, from radioactive sources as different as possible in regard to uranium and thorium content and geological horizon, should be re-examined, it by no means necessarily follows that any difference in atomic weight will be found.[88]

The first experimental results to be published on the atomic weight of lead of radioactive origin were read before the Chemical Society (London) on May 7, 1914 by Soddy. His collaborator was Henry Hyman, Mackay Smith Scholar at the University of Glasgow.[89] An abstract of their work appeared in the London *Morning Post* on the following day and in the Society's *Proceedings* on May 18, 1914.[90] From a kilogram of Ceylon thorite, a hydrated silicate of thorium with Th = 55%, U = 1 – 2%, and Pb = 0.4%, they isolated 1.2 grams of carefully purified lead chloride. They estimated the atomic weight of lead relative to that of ordinary lead (purified in an identical series of operations) volumetrically by titration with the same silver nitrate solution. The lead chloride was weighed according to the method of Baxter and Wilson.[91] Soddy and Hyman obtained the value of 208.4 for the atomic weight of the thorite lead, calculated from that of the ordinary lead taken as 207.1. This was significantly higher than that of ordinary lead and only a little higher than the value of 208.2, which they estimated from the uranium:thorium ratio in the mineral. The estimation of the atomic weight was purely relative, and Soddy and Hyman concluded,

> Although not yet sufficient to establish finally the point at issue, these preliminary results show clearly a difference in the atomic weights in the expected direction of the right order of magnitude.[92]

The next paper to be published was by Maurice Curie, the nephew of Madame Curie. On June 8, 1914 his results[93] were presented to the Académie des Sciences in Paris by Charles Moureu, and on the following day a story describing Curie's report appeared in a local newspaper.[94] Curie's purpose was:

> to obtain experimental proof that such differences [between atomic weights for lead of different origins] actually exist: thus the theoretical predictions related to the classification and transformation of radioelements could be confirmed; furthermore, the conclusions could be extended to inactive elements other than lead, of which some could also be considered as mixtures of several substances of very close atomic weights and of nearly identical chemical properties.[95]

By employing the method of Jean Servais Stas, which involved purifying and separating the metal as lead and converting it to and weighing it as lead nitrate, a method requiring less manipulation than the more accurate silver method, Curie obtained atomic weights for lead samples from pitchblende (206.64), carnotite (206.36), yttro-tantalite (206.54) – uranium minerals containing negligible quantities of thorium –, monazite (207.08) – a thorium mineral containing relatively little uranium –, and galena (207.01) – the sulfide mineral of common lead. The atomic weights from uranium minerals were thus distinctly lower than that from common lead, while that from the thorium mineral was slightly higher, in agreement with the predictions of the group displacement laws. In each case the lead was purified by a series of operations until it showed the same atomic weight after a fresh purification.

The next report to be published (in *Comptes rendus* for June 15, 1914) on the atomic weight of lead of radioactive origin was the first one to be prepared by a recognized authority on atomic weight determinations, Otto Hönigschmid (1878–1945),[96–99] Professor of Chemistry and Director of the Laboratory for Inorganic and Analytical Chemistry at the Technische Hochschule at Prague (1911–1918). In 1909–1910 Hönigschmid spent a year at Harvard University, Cambridge, Massachusetts where he studied under Theodore William Richards, the acknowledged world master in the area of exact determination of atomic weights. Since 1911 Hönigschmid was associated with the Radium Institute, which had just opened in Vienna, and in 1918 he became Head of the Analytical Chemistry Department at the University of Munich, where he established his famous atomic weight laboratory, his primary interest for the rest of his life. After Wilhelm Ostwald's retirement in 1922, Hönigschmid became Chairman of the German Atomic Weight Commission, and he was a member of the International Committee on Atomic Weights since its creation in 1930.

Together with his student Mlle. Stephanie Horovitz, Hönigschmid[100] accurately determined the atomic weight of lead by the two methods used by Gregory Paul Baxter, another student of Richards', for the determination of the atomic weight accepted at the time. Both involved calculation of the ratio $PbCl_2$:2AgCl; one was exclusively gravimetric, while the other involved the use of a nephelometer. The purified lead chloride was isolated from residues obtained from treating pitchblende for the preparation of radium salts and provided by the Akademie der Wissenschaften zu Wien.

Hönigschmid had previously determined the atomic weight of radium (225.97)[101] and from it calculated that the atomic weight of radium G (lead), formed by loss of 5 alpha particles should be 205.62 ($225.97 - 5 \times 4.07$). He reasoned,

> Since the purest pitchblende, with 60% $U_3O_8$, also contains 2 to 3% lead,it is possible that at least a part of this lead would be radium G; the atomic weight of lead extracted from pitchblende should be below that of ordinary lead by an amount corresponding to its content of radium G.[102]

Hönigschmid and Horovitz obtained the value 206.736, or 0.4 below the atomic weight of ordinary lead, as the mean of six determinations varying over the extremes of 0.03 unit and of three determinations varying over 0.018 unit. They remarked:

It is possible that a lead with still lower atomic weight could be extracted from pitchblende free from isolated fragments of blende.[103]

Although Hönigschmid and Horovitz's paper was published on June 15, 1914, a week after Curie's paper, their results had been presented to the Bunsen Gesellschaft in Leipzig on May 23 and to the Deutsche Chemische Gesellschaft in Berlin. It was at the Leipzig meeting that Fajans also reported the results of Richards and Lembert, which we will now discuss.

Fajans had tried to make the atomic weight determinations himself but soon realized that the task required such precision that only the results of recognized authorities would be accepted widely. Indeed, the results of Soddy and Hyman, the first to be published, were regarded with skepticism.[104] In the Fall of 1913 Fajans and the Grossherzogliche Technische Hochschule, Karlsruhe, with the support of Professor Georg Bredig, sent Dipl. Ing. Max Ernst Lembert (1891–1925), a young *Doktorand* of Fajans', to the Wolcott Gibbs Memorial Laboratory at Harvard University to study technique under Theodore William Richards (1868–1928)[105–114] and to determine whether the atomic weight of lead from uranium does indeed differ from that of ordinary lead. At about the same time Sir William Ramsay urged, on behalf of Soddy, that Richards should make such a determination.

Richards, whose mentor at Harvard, Josiah P. Cooke, had trained him in atomic weight determinations, was then at the height of his fame. He was perhaps the only American scientist of his time to be offered a professorship from a German university (Göttingen, which he declined), and he was elected President of both the American Chemical Society (1912) and of the American Association for the Advancement of Science (1917). In the year that he carried out his atomic weight determination with Lembert he received the Nobel Prize in chemistry 'in recognition of his accurate determinations of the atomic weights of a large number of chemical elements',[115] a topic to which he had devoted almost his entire scientific career.[116–119] He was the first American chemist to be so honored, and he remained the only American Nobel laureate in chemistry until 1932 when Irving Langmuir received the prize 'for his discoveries and investigations in surface chemistry'. Either alone or with the aid of his students, Richards investigated the atomic weights of twenty-five elements, and if the atomic weights determined independently by his students Gregory Paul Baxter (1876–1953)[120] and Otto Hönigschmid are included, the number rises to fifty-five of the elements then known.

Richards and Lembert began their paper with statements relating their work both to the group displacement laws and to the concept of isotopes:

> It has been pointed out by many of the workers upon radioactivity, especially by Boltwood, Ramsay, Rutherford and Fajans, that the most conclusive test concerning the recent theory of the degeneration of radioactive elements is to be found in the determination of the atomic weights.... ordinary lead, radium B, D, and G, and perhaps, also at least one other radioactive product from thorium and one from actinium ... should have identical spectra and be inseparable by chemical means, but, coming from different sources, they should have different atomic weights.... The problem is one capable of a decisive gravimetric test; specimens of lead, consisting of different mixtures, obtained from different sources, should have different atomic weights.[121]

Like Soddy and Hyman, Richards and Lembert used the method of Baxter and Wilson. They purified the lead chloride by recrystallization in quartz and platinum vessels, followed by drying in a desiccator and fusion in a stream of hydrogen chloride and nitrogen in the quartz tube of the bottling apparatus previously described by Richards.[122] They tested the efficiency of their crystallization procedure by deliberately contaminating the lead with about one tenth its weight of bismuth, the impurity in the radioactive material that they feared the most. Four crystallizations removed any trace of bismuth that could be detected by qualitative means. The lead chloride was dissolved in water, the chlorine was precipitated by silver nitrate, and atomic weights were calculated from the ratios $2Ag:PbCl_2$ and $PbCl_2:2AgCl$. Control analyses giving the atomic weight of ordinary lead were carried out in exactly the same way and gave essentially the same value as that obtained by Baxter and Wilson[123] and by Baxter and Grover[124] in what was then unpublished work.

The sources of the samples, the persons who furnished them, and the final values for the atomic weights were as follows: (1) Uraninite, North Carolina, U.S.A., 'perhaps the most valuable of all our samples (because it came from very pure ore) ... practically of pure radioactive origin, no other lead except that from uraninite itself being included;' Bertram B. Boltwood, New Haven, Connecticut, U.S.A., and Ellen Gleditsch of Christiania (now Oslo), Norway, then working with

Boltwood; 206.40; (2) Pitchblende, Joachimstahl, Bohemia; Kasimir Fajans; 206.57; (3) Carnotite, Colorado, U.S.A.; Kasimir Fajans, obtained from Friedrich O. Giesel, Braunschweig, Germany; 206.59; (4) Thorianite, Ceylon; Bertram B. Boltwood and H. S. Miner, Chief Chemist, Welsbach Light Co., Gloucester City, New Jersey, U.S.A.; 206.82; (5) Pitchblende, Cornwall, England; Sir William Ramsay; 206.86; (6) Common lead from commercial lead acetate, Germany; 207.15. All of the radioactive samples thus gave a lower atomic weight than ordinary lead, as determined under identical conditions, the difference in one case – (1) above – amounting to as much as 0.75 unit.

With Gregory P. Baxter's help, the ultraviolet spectrum of a typical radioactive sample was compared with a sample of Baxter's purest ordinary lead in Baxter's Féry quartz spectrograph and found to be identical 'except for a trace of copper too small to affect the result, and a negligible trace of silver known to have been present'. Richards and Lembert referred to their results as 'striking' and 'amazing'. They concluded,

There can be no question that the radioactive samples contain another element having an atomic weight so much lower than that of ordinary lead as to admit of no explanation through analytical error, and yet so nearly like ordinary lead as not to have been separated from it by any of the rather elaborate processes to which we had subjected the various samples.[125] ... No simple linear quantitative relationship between the exact amount of radioactivity and the atomic weight of lead was found.... The inference seems to be that radioactive lead contains an admixture of some substance different from ordinary lead, and very difficult to separate from it by chemical means. This substance [Richards and Lembert did not use the term 'isotope'] cannot be identified in the ultraviolet spectrum of the material, either because it has the same spectrum as lead, or because it has no spectrum in that part of the field, or because its spectrum is masked or aborted by that of lead[126]...: It would perhaps be premature to indulge in further hypothetical reasoning concerning the nature of this extraordinary phenomenon, but the nature of the variation unquestionably points in the direction of the hypothesis of Fajans and of Soddy.[127]

As we have noted, Richards devoted the major part of his career to careful determinations of atomic weights, which were expected to be constant values. At first sight, the new results would thus seem to minimize the importance of atomic weights in general, but Richards and Lembert rejected this superficial view:

If the results which we have obtained really indicate that several kinds of lead having the same properties and spectrum may be mixed together and not separated chemically, it is evident the atomic weight becomes almost the only criterion, except radioactivity, capable of detecting the admixture and tracing the factors to their source.... Thus the idea that the supposed constants may possibly be variable, adds to the interest which one may reasonably take in their accurate determination, and enlarges the possible field of investigation instead of contracting it.[128]

They pointed out that results with lead were contrary to previous experience at Harvard with a number of other elements, especially copper (1887), calcium (1902), silver (1905), sodium (1905), chlorine (1905), and terrestrial and extra-terrestrial iron (1911), each of which was found to yield a constant atomic weight, regardless of the geographical source.

Richards and Lembert's article was submitted to the *Journal of the American Chemical Society* on May 14, 1914 (received May 16) and was published in the July, 1914 issue of the journal, making it the last of the four papers on the topic to appear in 1914. However, a brief announcement of their work was made by Fajans at the Leipzig meeting of the Bunsen Gesellschaft on May 21, 1914,[129] and a report of their work appeared in the June 5, 1914 issue of *Science*.[130] Thus, within about a single month the existence of isotopes was confirmed experimentally in four independent laboratories in four different countries – a tribute to the international character of science.

Richards and Lembert's work has been called 'the final act of the dramatic events surrounding the clarification of the place of the radio-elements in the periodic table'.[131] With their experimental confirmation of the existence of elements of the same atomic number but different atomic weights, a new dimension was added to Mendeleev's system. As time passed, more and more isotope mixtures were discovered; even the simplest element, hydrogen, was found to possess a heavier isotope – deuterium – for the discovery of which the American chemist Harold Clayton Urey was awarded the Nobel Prize in chemistry in 1934. Nevertheless, until Aston's development of the mass spectrograph in 1919 and his proof of the isotopes of neon,[132] the only conclusive evidence for isotopes lay in Richards and Lembert's definitive demonstration of the difference in atomic weight between ordinary lead and lead of radioactive origin.

## 5. Later Work on the Atomic Weight of Lead

Although Richards and Lembert's work was widely recognized as a confirmation of the existence of isotopes, Richards regarded their paper as 'only a preliminary one', and he and his students continued to work on what he called 'the problem of radioactive lead'.[133] Space limitations do not permit detailed examination of investigations by Richards, Hönigschmid, and others, but brief treatments of this work on the atomic weight of both ordinary lead and lead of radioactive origin are found in standard reference works.[134] We may, however, mention several highlights of the work closely following the classic research of 1914.

Since the work of Baxter and Wilson in 1907 on the atomic weight of lead, which yielded the value of 207.09, the atomic weight of this element was for a number of years under almost continuous investigation at Harvard University.[135] After Richards and Lembert's work, the question arose as to whether ordinary lead always possesses the same atomic weight. Therefore Richards' former student, Gregory Paul Baxter, together with Fred Leslie Grover, prepared carefully purified lead chloride and lead bromide from ores (either pure galena, PbS, or cerussite, $PbCO_3$) obtained from the United States, Germany, and New South Wales. They avoided radioactive minerals in order to eliminate uncertainty from the possible presence of radioactive lead. From eleven specimens, they obtained a value of 207.20, finding no evidence 'that common lead is composed either wholly or in part of isotopes of different atomic weight'.[136] Together with his students, Baxter continued atomic weight determinations of ordinary and 'radioactive' lead through the late 1930's.

Of the two varieties of lead of radioactive origin then known – 'uranium-lead' and 'thorium-lead' –, the former was more extensively investigated because thorium-free uranium minerals were more readily available than uranium-free thorium minerals. In 1915 Hönigschmid and Horovitz determined the atomic weight of 'uranium-lead' obtained from pure crystalline pitchblende from Morogoro, East Africa (206.046) and from bröggerite from Moss, Norway (206.063).[137] In the following year, Richards and Wadsworth[138] confirmed the earlier results of Richards and Lembert, using new samples of varied origin. Their results may be summarized: (1) Galena (common lead); U.S.A.; 207.18; (2) Carnotite; Colorado, U.S.A.; 207.00; (3) Carnotite; Radium Hill, New South Wales, Australia; 206.34; (4) Bröggerite, Moss, Norway; 206.12; (5) Cleveite; Langesund, Norway; 206.08. They regarded the fact that the most carefully selected sample – (5) – gave the lowest result as strong evidence that the higher results obtained from other samples were due merely to the accidental admixture of ordinary lead.

Richards[139] compared his lowest value (206.08) with the atomic weight of 'uranium-lead' calculated according to the group displacement laws as follows:

| | | | |
|---|---|---|---|
| Atomic weight of uranium | 238.18 | | |
| 8 × atomic weight of helium | 32.00 | | |
| Residue (lead?) | 206.18 | = | 206.18 |
| Atomic weight of radium | 225.96 | | |
| 5 × atomic weight of helium | 20.00 | | |
| Residue (lead?) | 205.96 | = | 205.96 |
| Average hypothetical value for lead | | = | 206.07 |
| Observed value for uranium-lead | | = | 206.08 |
| Difference | | | 0.01 |

Not given to hyperbole, Richards nevertheless regarded the remarkably close agreement as 'a striking confirmation of the theory' and the calculations as 'perhaps the most successful attempt on record to compute an atomic weight from hypothetical assumptions'.

Richards did not limit his investigations to atomic weight determinations but painstakingly searched for other differences between 'radioactive' lead and the common metal. He investigated their densities and found them to be very nearly proportional to their atomic weights, confirming the assumption that the atomic volumes are almost exactly the same although the atomic weights are so markedly different.[140] He found 'the spectrum lines produced by these two sorts of lead, when heated to the high temperature of the electric arc, are so precisely alike, both as to their wave-lengths and their intensities, that no ordinary spectrum analysis shows any difference whatever',[141] an observation disputed by others.[142] Richards and Schumb[143] also showed that the refractive index of common lead nitrate is identical with that of the nitrate of 'radioactive' lead to within 1 part in nearly 20 000, demonstrating than density is a less important factor in determining refractive index that had been assumed. They also found the molal solubilities of the two salts to be virtually identical, a not unexpected result considering the impossibility of separating the two isotopes of lead by fractional crystallization. Richards and Hall[144] carried out one thousand such crystallizations but failed to detect any difference in atomic weight between the extreme fraction of the crystals and the extreme fraction of the mother liquor. Richards and his students also failed to detect any thermoelectric effect between wires of the two types of lead, which were also found to have identical melting points.

The question of thorium-lead was not conclusively settled until the 1920's or even later, largely because of the lack of uranium-free thorium minerals. In 1915 Soddy[145] isolated metallic lead from thorite and from the ratio Pb:$PbCl_2$ found its atomic weight to be 207.694 as compared to 207.199 for ordinary lead, determined in the same way. The research was interrupted by the advent of World War I. Hönigschmid,[146] however, used another portion of Soddy's thorite lead and by the silver titration method obtained the value 207.77 ± 0.014. Hönigschmid's highest value, 207.9, was so much above the atomic weight of ordinary lead that the reality of thorium-lead could no longer be doubted.[147]

*Department of Chemistry*
*California State University, Fresno*
*Fresno, CA 93740, U.S.A.*

## Notes and Bibliography

* Reprinted from the *Journal of Chemical Education* **59**, 1 (January 1982): 3–8; **59**, 2 (February 1982): 119–123.

[1] Russell, Alexander S.: 1956, 'Prof. Frederick Soddy, L.L.D., F.R.S.', *Chemistry and Industry* 1420–21.

[2] Kent, Andrew: 1963, 'Frederick Soddy (1877–1956)', *Proceedings of the Chemical Society* 327–30.

[3] Fleck, Alexander: 1957, 'Frederick Soddy', *Biographical Memoirs of Fellows of the Royal Society* **iii**, 203–16.

[4] Howorth, Muriel: 1958, *Pioneer Research on the Atom: Rutherford and Soddy in a Glorious Chapter of Science: The Life Story of Frederick Soddy*, London.

[5] Trenn, Thaddeus J.: 1975, 'Frederick Soddy', in Charles Coulston Gillispie (ed.), *Dictionary of Scientific Biography*, New York, **xii**, 504–509. This article contains numerous references to articles by and about Soddy.

[6] Rutherford, E. and Soddy, Frederick: (a) 1902, 'The Radioactivity of Thorium Compounds: I. An Investigation of the Radioactive Emanation', *Journal of the Chemical Society (London)* **Lxxxi**, 321–50; (b) 1902, 'idem: II. The Cause and Nature of Radioactivity', *J. Chem. Soc.*, **Lxxxi**, 837–60; reprinted in *Phil. Mag.* [6] **iv** (1902), 370–396; (c) 1903, 'Radioactive Change', *Phil. Mag.*, [6] **V**, 576–91. Papers (a), (b), and (c) have been reprinted in Alfred Romer, 1964, *The Discovery of Radioactivity and Transmutation*, New York, 87–116, 124–50, and 152–66, respectively. For an historical evaluation of Rutherford and Soddy's work and that of other early workers on radioactive elements see G. E.

M. Jauncey, 1946, 'The Early Years of Radioactivity', *American Journal of Physics* **xiv**, 226–41.

[7] Ramsay, William and Soddy, Frederick: 1903, *Proceedings of the Royal Society* **Lxxii**, 204; 'Gases Occluded by Radium Bromide', *Nature* **Lxviii** (1903), 246; reprinted in Romer, *Discovery*, 171–2.

[8] Soddy, Frederick: (a) 1923, 'The Origins of the Conceptions of Isotopes', in *Les Prix Nobel en 1921–1922*, Stockholm, 1–29; (b) also in 1966, Nobel Foundation, *Nobel Lectures, Including Presentation Speeches and Laureates' Biographies, Chemistry 1901–1921*, Amsterdam, London, New York, 371–99.

[9] Hahn, O.: 1905, 'Über ein neues die Emanation des Thoriums gebendes radioaktives Element', *Jahrbuch der Radioaktivität und Elektronik* **ii**, 233–66; 'A New Radioactive Element which Emits Thorium Emanation', *Chemical News* **xcii** (1905), 251–2; also in *Proc. Roy. Soc.* **Lxxvi** (1905), 115–7; reprinted in Alfred Romer, *Radiochemistry and the Discovery of Isotopes*, New York 1970, 107–11.

[10] Elster, J. and Geitel, H.: 1906, 'Beiträge zur Kenntnis der Radioaktität des Thoriums', *Physikalische Zeitschrift* **vii**, 445–52.

[11] Blanc, G. A.: 1905, 'Über die Natur der radioaktiven Elemente, welche in den Sedimenten der Thermalquellen von Echaillon und von Salins-Moutiers (Savoyen) enthalten sind', *Physik. Z.* **vi**, 703–7; 'Untersuchungen über ein neues Element mit den radioaktiven Eigenschaften des Thors', *ibid.* **vii** (1906), 620–30.

[12] Boltwood, B. B.: 1906, 'Activity of Thorium Minerals and Salts', *American Journal of Science* [4] **xxi**, 415–26.

[13] Dadourian, H. M.: 1906, 'The Radioactivity of Thorium', *Amer. J. Sci.* [4] **xxi**, 427–32.

[14] McCoy, Herbert N. and Ross, W. H.: 1906, 'The Relation between the Radioactivity and the Composition of Thorium Compounds', *Amer. J. Sci.* [4] **xxi**, 433–43.

[15] Hahn, O.: 1907, 'Ein neues Zwischenprodukt im Thorium, *Physik. Z.* **viii**, 277–81; *Berichte der Deutschen Chemischen Gesellschaft* **XL** (1907), 1462–9. For an English translation see Romer, *Radiochemistry*, 112–9.

[16] Hahn, Otto: 1908, 'Ein kurzlebiges Zwischenprodukt zwischen Mesothor und Radiothor', *Physik. Z.* **ix**, 246–8; for an English translation see Romer, *Radiochemistry*, 139–44.

[17] McCoy, Herbert N. and Ross, W. H.: 1907, 'The Specific Radioactivity of Thorium and the Variation of the Activity with Chemical Treatment and with Time', *Journal of the American Chemical Society* **xxix**, 1709–18; reprinted in Romer, *Radiochemistry*, 128–39.

[18] Stieglitz, Julius: July 10, 1935, 'Herbert Newby McCoy', *Industrial and Engineering Chemistry, News Edition* **xiii**, 280.

[19] 'Award of the Willard Gibbs Medal to H. N. McCoy,' *Chemical Bulletin* **xxiv** (1937), 207–24.

[20] Footnote 17, 1711.

[21] Footnote 8b, 376.

[22] *Ibid.*, 397.

[23] Footnote 21.

[24] Soddy, Frederick: 1908, 'Radioactivity', *Annual Progress Report to the Chemical Society for 1907* **iv**, 326. Soddy's reports, dating from 1904 to 1920 and containing much otherwise unpublished work on isotopes, have been published in facsimile and edited with commentary by Thaddeus J. Trenn, *Radioactivity and Atomic Theory*, London, 1975.

[25] Boltwood, Bertram B.: 1908, 'On Ionium, a New Radio-Active Element', *Amer. J.*

*Sci.*, [4] **xxv**, 365–81; reprinted in Romer, *Radiochemistry*, 147–63.

26 Marckwald, W. and Keetman, B.: 1908, 'Notiz über das Ionium', *Ber.* **xLi**, 49–50.

27 Keetman, Bruno: 1909, 'Über Ionium', *Jahrb. Radioakt. Elekt.* **vi**, 265–74; extracted from the 1909 doctoral dissertation of the author, who was a graduate student of Willy Marckwald at the University of Berlin; for an English translation see Romer, *Radiochemistry*, 164–73.

28 von Welsbach, Carl Auer: 1910, 'Mitteilungen der Radium Kommission, VI', *Sitzungsberichte der Kaiserlichen Akademie der Wissenschaften zu Wien*, **Cxix**, ii, a, 1.

29 Soddy, Frederick: 1911, 'The Chemistry of Mesothorium', *J. Chem. Soc.* **xCix**, 72–83 (82); reprinted in Romer, *Radiochemistry*, 179–90.

30 Herchfinkel, H.: 1910, 'Sur le radioplomb,' *Le Radium*, *Paris* **vii**, 198–200.

31 Paneth, Fritz A. and von Hevesy, Georg: 1913, 'Über Versuche zur Trennung des Radium D von Blei', *Monatshefte für Chemie* **xxxiv**, 1393–1400; 'Über Radioelementen, als Indikatoren in der analytischen Chemie, *ibid.*, 1401–8.

32 von Hevesy, G.: 1923, 'The Absorption and Translocation of Lead by Plants. A Contribution to the Application of the Method of Radioactive Indicators in the Investigation of the Change of Substance in Plants', *The Biochemical Journal* **xvii**, 439–45; reprinted in part in Henry M. Leicester (ed.), *Source Book in Chemistry 1900–1950*, Cambridge, Mass., 1968, 49–50.

33 Strömholm, D. and Svedberg, The: 1909, (a) 'Untersuchungen über die Chemie der radioaktiven Grundstoffe. I,' *Zeitschrift für anorganische Chemie*, **Lxi**, 338–48; (b) 'Idem, II', *ibid.*, **Lxiii** (1909), 197–206.

34 Godlewski, T.: 1905, 'A New Radioactive Product from Actinium', *Nature* **Lxxi**, 294–5; 'Actinium and its Successive Products', *Phil. Mag.* [6] **X** (1905), 35–45.

35 Footnote 8b, 381.

36 *Ibid.*

37 Footnote 33b.

38 Moseley, H. G. J. and Darwin, C. G.: 1913, 'The Reflexion of X Rays,' *Phil. Mag.* **xxvi**, 210–32; Moseley, H. G. J., 'The High-Frequency Spectra of the Elements', *ibid.*, **xxvi** (1913), 1025–34; **xxvii** (1914), 703–13.

39 Marckwald, W.: 1910, 'Zur Kenntnis des Mesothoriums', *Ber.* **xLiii**, 3420–2; for an English translation see Romer, *Radiochemistry*, 175–7.

40 Footnote 29.

41 Soddy, Frederick: 1911, 'Radioactivity', *Ann. Rep. 1910* **vii**, 285.

42 Fleck, footnote 3, 208.

43 Footnote 41, 286.

44 Thomson, J. J.: 1913, 'Bakerian Lecture', *Proc. Roy. Soc.* **Lxxxix**, A, 1; 'Some Further Applications of the Method of Positive Rays', *Nature* **xci** (1913), 333–7; 'Positive Rays of Electricity', *ibid.*, 362.

45 Aston, Francis W.: 1913, paper communicated to the British Association for the Advancement of Science, Section A, Birmingham, 'New Element in the Atmosphere', *Engineering* **xcvi**, 423.

46 Soddy, Frederick: 1913, 'Radioactivity', *Ann. Rep. 1912* **x**, 265–6.

47 Fleck, Alexander: 1913, 'The Chemical Nature of Some Radioactive Disintegration Products', *J. Chem. Soc.* **ciii**, 381–99; 'idem. Part II', *ibid.*, **ciii** (1913), 1052–61. The first article is reprinted in Romer, *Radiochemistry*, 228–46.

48 *ἴσος*, same; *τόπος*, place. For details on the isotopes of various elements see F. W.

Aston: 1933, *Mass-Spectra and Isotopes*, London, 1933.
[49] (a) Soddy, Frederick: Dec. 4, 1913, 'Intra-atomic Charge', *Nature* **xcii**, 399–400; (b) reprinted in 'Isotopes – A Fiftieth Anniversary', *Proc. Chem. Soc.* (1963), 325–26 and in (c) Romer, *Radiochemistry*, 250–52.
[50] Soddy, Frederick: 1921, 'Contribution to a Discussion on Isotopes', *Proc. Roy. Soc.* **xcix**, 98.
[51] Fleck, Alexander: 1963, 'Early Work in the Radioactive Elements', *Proc. Chem. Soc.*, 330.
[52] Preyer, W.: 1893, *Das genetische System der chemischen Elemente*, Berlin.
[53] Fajans, K.: 1913, 'Die radioaktiven Umwandlungen und die Valenzfrage vom Standpunkte der Struktur der Atome', *Verhandlungen der Deutschen Physikalischen Gesellschaft* **xv**, 240–59.
[54] Fajans, K.: 1919, *Radioaktivität und die neueste Entwickelung der Lehre von den chemischen Elementen*, Braunschweig, 35.
[55] Footnote 49b, 326.
[56] Soddy, Frederick: 1909, 'Multiple Atomic Disintegration: A Suggestion in Radioactive Theory', *Phil. Mag.* **xviii**, 739; 'Multiple Disintegration', *Annual Report 1911*, **ix** (1912), 311–6.
[57] Cranston, J. A.: 1963, 'The Group Displacement Law', *Proc. Chem. Soc.*, 330–1.
[58] Anders, Oswald U.: 1964, 'The Place of Isotopes in the Periodic Table: The 50th Anniversary of the Fajans-Soddy Displacement Laws', *Journal of Chemical Education* **xLi**, 522–5.
[59] Footnotes 29 and 47.
[60] von Lerch, F.: 1906, 'Trennung des Radiums C vom Radium B', *Annalen der Physik* **xx**, 345–54.
[61] Lucas, Richard: 1906, 'Über das elektrochemische Verhalten der radioaktiven Elemente', *Phys. Z.* **vii**, 340–2.
[62] Cameron, A. T.: 1909, 'The Position of the Radio-Active Elements in the Periodic Table', *Nature* **Lxxxii**, 67–8.
[63] Soddy, Frederick: 1911, *The Chemistry of the Radioelements*, London, 29.
[64] Russell, A. S.: 1913, 'The Periodic System and the Radio-Elements', *Chem. News*, **cvii**, 49–52.
[65] Fajans, K.: 1912, *Le Radium* **ix**, 239.
[66] von Hevesy, G.: 1912, 'Über den Zusammenhang zwischen den chemischen Eigenschaften der Radioelemente und der Reihenfolge radioaktiver Umwandlungen', *Phys. Z.* **xiii**, 672–3.
[67] Badash, Lawrence: 1979, 'The Suicidal Success of Radiochemistry', *British Journal for the History of Science* **xii**, 245–56.
[68] Footnote 63.
[69] An adumbration of isotopes and the displacement laws is found in A. T. Cameron, *Radiochemistry*, London, 1910, 141.
[70] Fajans, K.: Dec., 1912, *Verhandlungen der Naturhistorisch-Medizinischer Verein, Heidelberg* **xii**, 173.
[71] Fajans, K.: 1913, (a) 'Über eine Beziehung zwischen der Art einer radioaktiven Umwandlung und dem Elektrochemischen Verhalten der betreffenden Radioelemente', *Physik. Z.* **xiv**, 131–6; (b) 'Die Stellung der Radioelemente im periodischen System',

*ibid.*, 136–42. For English translations of (a) and (b), see Romer, *Radiochemistry*, 198–206 and 207–18, respectively.
[72] Footnote 71b.
[73] Footnote 7.
[74] Rutherford, E.: 1906, 'The Mass and Velocity of $\alpha$-Particles of Radium and Actinium', *Phil. Mag.* [6] **xii**, 348–71.
[75] Footnote 54.
[76] Footnote 49.
[77] Fajans, Kasimir and Göhring, Oswald: 1913, 'Über die komplexe Natur des Ur X', *Naturwissenschaften* **i**, 339; 'Über das Uran $X_2$ – das neue Element der Uranreihe', *Physik. Z.* **xiv** (1913), 877–84.
[78] Soddy, Frederick: 1913, 'Origin of actinium', *Nature* **xci**, 634–5.
[79] Soddy, Frederick and Cranston, J. A.: 1918, 'The Parent of Actinium', *Nature* **c**, 498–9; *Proc. Roy. Soc. (London)* **xcivA** (1918), 384.
[80] Hahn, Otto and Meitner, Lise: 1918, 'Die Muttersubstanz des Actiniums, ein neues radioaktives Element von langer Lebensdauer', *Physik. Z.* **xix**, 208–18.
[81] Aston, Francis W.: 1935, 'The Story of Isotopes', *British Association for the Advancement of Science Report*, Presidental Address to Section A, 26.
[82] Krivomazov, A. N.: 1976, 'Problemy prognozirovaniia na rannikh etapakh razvitiia ucheniia o radioaktivnosti', in *Prognozirovanie v uchenii o periodichnosti*, Moscow, 224–46 (239–47).
[83] In modern nuclear equations,

$$^{238.5}_{92}\mathrm{U} \rightarrow {}^{206.5}_{82}\mathrm{Pb} + 8^{4}_{2}\alpha + 6^{0}_{-1}\beta$$

$$^{232.4}_{90}\mathrm{Th} \rightarrow {}^{2084}_{82}\mathrm{Pb} + 6^{4}_{2}\alpha + 4^{0}_{-1}\beta$$

[84] Footnote 71a.
[85] Fajans, Kasimir: 1913, 'Die radioaktiven Umwandlungen und das periodische System der Elemente', *Ber.* **xLvi**, 422–39 (430); reprinted in part in Leicester, *Source Book*, 68–75.
[86] Soddy, Frederick: 1913, 'The Radio-Elements and the Periodic Law', *Chem. News* **cvii**, 97–9; reprinted in part in Leicester, *Source Book*, 75–8 and in full in Romer, *Radiochemistry*, 219–26.
[87] Soddy, Frederick: 1914, 'Radioactivity', *Ann. Rep. 1913* **x**, 270.
[88] *Ibid.*
[89] Soddy, Frederick and Hyman, Henry: 1914, 'The Atomic Weight of Lead from Ceylon Thorite', *J. Chem. Soc* **CV**, 1402–8.
[90] Soddy, Frederick and Hyman, Henry: 1914, 'The Atomic Weight of Lead from Ceylon Thorite', *Proc. Chem. Soc.* **xxx**, 134–5.
[91] Baxter, Gregory Paul and Wilson, John Hunt: 1908, 'A Revision of the Atomic Weight of Lead. Preliminary Paper – The Analysis of Lead Chloride', *J. Am. Chem. Soc.* **xxx**, 187–95; *Proceedings of the American Academy of Arts and Sciences* **xLiii** (1907), 363; 'Revision des Atomgewichtes von Blei, I. Mitteilung: Die Analyse von Bleichlorid', *Z. anorg. Chem.* **Lvii** (1908), 174–84.
[92] Footnote 89, 1407.
[93] Curie, Maurice: 1914, 'Sur les écarts de poids atomiques obtenus avec le plomb

provenant de divers minéraux', *Comptes rendus hebdomadaires des séances de l'Académie des Sciences, Paris* **cLviii**, 1676–79.

[94] 'Il'y a plusieurs espèces de plomb', *Le Matin*, Paris, June 9, 1914, p. 1.

[95] Footnote 93, 1677.

[96] Zintl, E.: 1938, 'Otto Hönigschmid zum 60. Geburtstag,' *Z. anorg. allgem. Chem.* **ccxxxvi**, 3–11.

[97] Oesper, Ralph E.: 1940, 'Otto Hönigschmid', *J. Chem. Educ.* **xvii**, 562.

[98] Birckenbach, L.: 1949, 'Otto Hönigschmid', *Chemische Berichte* **Lxxxii**, xi–Lxv.

[99] Ronge, Grete: 1972, 'Otto Hönigschmid', in Charles Coulston Gillispie (ed.), *Dictionary of Scientific Biography*, New York, Vol. VI, 480–1.

[100] Hönigschmid, O. and (Mlle.) St.-Horovitz: 1914, 'Sur le poids atomique du plomb de la pechblende', *Compt. rend.* **cLviii**, 1796–8. For an English translation see Leicester, *Source Book*, 82–4.

[101] Hönigschmid, O.: 1912, 'Revision des Atomgewichtes des Radiums und Herstellung von Radiumstandardpräparaten', *Monatsh. Chem.* **xxxiii**, 253–88.

[102] Footnote 100, 1797.

[103] Footnote 100, 1798.

[104] Fajans, K., conversation with Lawrence Badash, July 16–17, 1966; K. Fajans, letter to James B. Conant, October 30, 1969; T. W. Richards, letter to Bertram B. Boltwood, January 10, 1977.

[105] Harrow, Benjamin: 1920, 'Richards and Atomic Weights', *Eminent Chemists of Our Time*, New York, Chapter 4.

[106] Baxter, Gregory P.: 1928, 'Theodore William Richards', *Science* **Lxviii** 333–9.

[107] Briscoe, H. V. A.: 1928, 'Prof. T. W. Richards, Foreign Member of the Royal Society', *Nature* **cxxii**, 28–9.

[108] Hartley, Harold: 1930, 'Theodore William Richards Memorial Lecture', *J. Chem. Soc.*, 1937–69.

[109] Forbes, George Shannon: 1932, 'Investigations of Atomic Weights by Theodore William Richards', *J. Chem. Educ.* **ix**, 453–8.

[110] Ihde, Aaron J.: 1961, 'Richards – Corrector of Atomic Weights, 1868–1928', in Eduard Farber (ed.), *Great Chemists*, New York, 822–6.

[111] Ihde, Aaron J.: 1969, 'Theodore William Richards and the Atomic Weight Problem', *Science* **cLxiv**, 647–51.

[112] Kopperl, Sheldon J.: 1970, *The Scientific Work of Theodore William Richards*, Dissertation, University of Wisconsin, Madison.

[113] Conant, James B.: 1970, 'Theodore William Richards and the Periodic Table', *Science* **cLxviii**, 425–6.

[114] Kopperl, Sheldon J.: 1975, 'Theodore William Richards', in Charles Coulston Gillispie (ed.), *Dictionary of Scientific Biography*, New York, Vol. xi, 416–8.

[115] Richards, Theodore W.: 1966, 'Atomic Weights', in Nobel Foundation, *Nobel Lectures, Including Presentation Speeches and Laureates' Biographies, Chemistry 1901–1921*, Amsterdam, London, New York, 280–92.

[116] Richards, Theodore William: 1910, *Determination of Atomic Weights*, Washington.

[117] Richards, Theodore William: 1915, 'Recent Researches in the Wolcott Gibbs Memorial Laboratory of Harvard University', *Science* **xLii**, 845–51.

[118] Richards, Theodore William: 1924, 'Atomic Weights and Isotopes', *Chemical Reviews* **i**, 1–40.

[119] Richards, Theodore William: 1916, 'Ideals of Chemical Investigation', *Science* **xLiv**, 37–45; *Harvard Graduates' Magazine* **xxv** (1916), 1–10; *J. Chem. Educ.* **vi** (1929), 2239–45.
[120] MacNevin, William Marshall: 1954, 'Experiences in the Laboratory of Gregory Paul Baxter', *J. Chem. Educ.* **xxxi**, 303–5; Arthur Vyvyan Howard, 'Gregory Paul Baxter', in *Chambers's Dictionary of Scientists*, New York, 1958, 43.
[121] Richards, Theodore W. and Lembert, Max E.: 1914, 'The Atomic Weight of Lead of Radioactive Origin', *J. Am. Chem. Soc.* **xxxvi**, 1329–44; reprinted in part in Leicester, *Source Book*, 79–82. For a French summary see 'Poids atomique du plomb d'origine radioactive,' *Compt. Rend.* **cLix** (1914), 248–50.
[122] Richards, Theodore William: 1911, 'The Fundamental Properties of the Elements (Faraday lecture of 1911)', *J. Chem. Soc.* **xcix**, 1201–18.
[123] Footnote 91.
[124] Baxter, Gregory Paul and Grover, Fred Leslie: 1915, 'A Revision of the Atomic Weight of Lead. The Analysis of Lead Bromide and Chloride [Third Paper]', *J. Am. Chem. Soc.* **xxxvii**, 1027–61.
[125] Footnote 121, 1331.
[126] Footnote 121, 1344.
[127] Footnote 121, 1343.
[128] Footnote 121, 1342–43.
[129] Fajans, K.: 1914, 'Nachtrag zu dem Aufsatz '"Die Radioelemente und das periodische System"', *Naturw.* **ii**, 543–44; 'Ueber die Endprodukte radioaktiver Zerfallsreihen', *Zeitschrift für Elektrochemie* **xx** (1914), 319, 449–52.
[130] 'The Atomic Weight of Lead of Radioactive Origin', *Science* **xxxix** (1914), 831–2.
[131] Footnote 58.
[132] Footnote 48.
[133] Richards, Theodore William: 1919, 'The Problem of Radioactive Lead', *Science* **xLix**, 1–11; reprinted in *Nature* **ciii** (1919), 74–8, 93–6 and in *Smithsonian Report for 1918*, (1920), 205–19.
[134] Mellor, J. W.: 1927, *A Comprehensive Treatise on Inorganic and Theoretical Chemistry*, New York, Vol. VII, 600–4; René Faivre and Raymond Weiss, 'Détermination de la mass atomique du plomb,' in Paul Pascal (ed.), *Nouveau traité de chimie minérale*, Paris, 1963, Vol. VIII, 472–5; *Gmelins Handbuch der anorganischen Chemie*, 8th ed., Weinheim/Bergstrasse, 1972, 'Blei', System Nr. 47, Part B1, 1–9.
[135] Baxter, Gregory Paul and Thorvaldson, Thorbergur: 1915, 'A Revision of the Atomic Weight of Lead. The Analysis of Lead Bromide [Second Paper]', *J. Am. Chem. Soc* **xxxvii**, 1020–27.
[136] Footnote 124.
[137] Hönigschmid, O. and Horovitz, S.: 1915, 'Über das Atomgewicht des "Uranbleis"', *Monatsh. Chem.* **xxxvi**, 355–80.
[138] Richards, Theodore William and Wadsworth, Charles, 3D: 1916, 'Further Study of the Atomic Weight of Lead of Radioactive Origin', *J. Am. Chem. Soc.* **xxxviii**, 2613–22.
[139] Footnote 133, 6–7.
[140] Richards, Theodore William and Wadsworth, Charles, 3D: 1916, 'The Density of Lead from Radioactive Minerals', *J. Am. Chem. Soc.* **xxxviii**, 221–7; 'Density of Radio-Lead from Pure Norwegian Cleveite', *ibid.* **xxxviii** (1916), 1658–60.

[141] Footnote 133, 7–8.
[142] Merton, Thomas R.: 1915, 'The Spectra of Ordinary Lead and Lead of Radioactive Origin', *Proc. Roy. Soc.* **xci A**, 198–201; William D. Harkins and Lester Aronberg, 'A Force, Apparently Due to Mass, Acting on an Electron, and the Non-Identity of Isotopes in Spectra and Other Properties', *J. Am. Chem. Soc.* **xLii** (1920), 1328–35.
[143] Richards, Theodore William and Schumb, Walter C.: 1918, 'The Refractive Index and Solubilities of the Nitrates of Lead Isotopes', *J. Am. Chem. Soc.* **XL**, 1403–9.
[144] Richards, Theodore William and Hall, Norris F.: 1917, 'Attempt to Separate the Isotopic Forms of Lead by Fractional Crystallization of the Nitrate', *J. Am. Chem. Soc.* **xxxix**, 531–41.
[145] Soddy, Frederick: 1915, 'The Density of Lead from Ceylon Thorite', *Nature* **xciv**, 615; 'The Atomic Weight of "Thorium" Lead', *ibid.*, **xcviii** (1916–1977), 469.
[146] Hönigschmid, Otto: 1917, 'Über das Thoriumblei', *Z. Elektrochem.* **xxiii**, 161–5; 'Neuere Atomgewichtsbestimmungen (Thoriumblei, Skandium)', *ibid.*, **xxv** (1919), 91–7.
[147] Kovarik, A. F. and McKeehan, L. W.: March 1929, 'Radioactivity', *Bulletin of the National Research Council*, **Li**, 25–8; Arthur Holmes, *The Age of the Earth*, London, 1927, 62; S. I. Levy, *The Rare Earths: Their Occurrence, Chemistry, and Technology*, London, 1924, 98–101.

JAN W. VAN SPRONSEN

# 5. SODDY AND THE CLASSIFICATION OF THE ELEMENTS

*The periodic system of the chemical elements was discovered during the 1860s on the basis of experiment rather than theory. In 1913 its theoretical confirmation as a classification system appeared from several directions by a replacement of the concept of atomic weight by that of atomic number, a modification to which Soddy contributed greatly. In the second part of his book* The Chemistry of the Radio-Elements, *subtitled* Radio-Elements and the Periodic Law *(1914), he published a three-dimensional helical representation of the periodic law, which successfully showed the central place of the noble gases and their function in linking the alkali metals with the halogens. Even earlier, with his concept of isotopy, Soddy clearly realized that most of the newly discovered products of the several radioactive disintegration series were not new elements which had to be positioned within the periodic system, and he defended such a classification against those who would discredit the periodic system altogether. With his displacement law of the elements (1913), the classification of all radioactive elements, including the end products of the radioactive series, received a practical and theoretical explanation as significant as Bohr's atomic theory of the same year.*

## 1. Introduction

The idea of classification is as old as mankind itself, for man seems to possess an innate desire to arrange things in some logical order.[1] In the animal kingdom, for instance, it was obvious to divide animals into those inhabiting the water, the earth, and the sky, i.e., a natural division into fishes, mammals, and birds. However, exceptions to such a classification can result in problems. For example, in the water there also live mammals, such as whales, as well as walking rather than swimming animals, such as crocodiles. Similar exceptions also had to be made for animals living on the earth and in the sky, and other refinements became necessary. In the course of world history such taxonomic changes were made a number of times, from Aristotle to Cuvier, to name only

*George B. Kauffman (ed.), Frederick Soddy (1877–1956)*, 93–112.

two of the most important scientists in this field of classification.

Classification in the vegetable and animal kingdoms exhibits the fundamental difference between natural and artificial classifications. An artificial classification cannot be a true one; it will change as more discoveries occur, as new insights in classification arise, or as other criteria are applied. More than one artificial classification can be devised, but only one natural classification, determined by nature or instituted by the Creator, can exist.

As far as the chemical elements are concerned, when only a few of them were known and before their exact nature was understood (up to the end of the eighteenth century), the need for their classification was not very great. It is not clear whether Robert Boyle's attempt at a classification considered the metals as elements or only their oxides as elementary substances, in accordance with the then dominant phlogiston theory. The chemical elements should not be classified as pure substances, i.e., metals or nonmetals, but their classification should be based on their chemical properties. Thus consideration of typical properties of the free elements themselves, such as state of aggregation, conductivity of heat or electricity, color, melting point, etc. – all physical properties – results only in artificial classifications.

The first modern scientist to search for a classification of the elements – Johann Wolfgang Döbereiner in 1817 – considered compounds of the elements, which would result in a chemical classification. However, in order to discover one and only one natural system, the number of known elements had to be great enough for the natural relationship between their properties to be discernable.

In 1817 this was not yet the case; it was not yet possible to discover such a system with only about forty elements known from among the roughly ninety natural elements. Yet during the first half of the nineteenth century it was possible to discover parts of the system, e.g., Döbereiner's triads. However, it was only when the number of elements known had been expanded to about sixty that scientists were able to discern the larger relationships between the elements. This took place only in the 1860s after the so-called horizontal relationships between elements had been discovered in 1857 by William Odling and Jean Baptiste Dumas and after the vertical relationships (triads and longer series of elements with analogous properties) were known. These relationships were found without recourse to any theoretical foundation or background, but one hint existed pertaining to the natural system of the chemical elements which was discovered during the 1860s by six scientists working entirely independently of one another, viz., Alexandre-Émile Béguyer de Chancourtois, John Alexander Reina Newlands, William Odling, Gustavus Detlef Hinrichs, Lothar Meyer, and Dmitriĭ Ivanovich Mendeleev.[2,3] This hint was the periodic recurrence of the chemical properties of the elements when they were arranged according to increasing atomic weights.

With sixty known elements it was possible to construct this periodic system in principle. The individual elements were found to be the natural building blocks of this system. Although additional elements were discovered later, all of them were eventually fitted into the system. Thus rock samples brought back to earth from the moon have not been found to contain any elements not occurring on the earth, and it is unlikely that any elements which do not fit into the periodic system will be found, even on other planets.

## 2. The Periodic System of the Chemical Elements

As mentioned above, in 1817 Döbereiner recognized the relationship between the chemical properties of a number of elements and their atomic weights, a quantity defined only a few years earlier by John Dalton and determined by him for a number of elements. Yet there were several reasons why the first periodic system was not discovered for nearly another half-century. First, during the first half of the nineteenth century, too few elements were known to form the basis for such a system. Also, the battle between partisans of atomic weights and those of equivalent weights of the elements was still raging. Furthermore, there was not yet a clear insight into the valences of the chemical elements, which was a necessary prerequisite for the construction of a periodic system. When these necessary conditions were satisfied – in the 1860s –, the periodic system was discovered. Before considering these discoveries, a definition for a suitable periodic system is in order.

This definition differs somewhat from the present-day definition in which atomic number is substituted for atomic weight. During the time of the discovery of the periodic system, nothing was known about the number of protons, the nuclear charge, and the atomic number. During the 1860s scientists based their systems on atomic weights, which by then had been determined unequivocally for most elements. At that time the periodic system was a classification in which all elements were arranged according to atomic weights and in which elements with analogous properties belonged to the same groups (rows or columns).

Each of the six discoverers arrived at his system on different grounds. The first discoverer, the Frenchman Béguyer de Chancourtois, was not even a chemist. As a mineralogist and geologist, he was strongly interested in classification. He even hoped to create a new language based on classification of letters and sounds. In 1862 he proposed a classification of the elements in which he constructed the first periodic system. Inspection of this classification discloses many deviations from the rule 'systematizing analogous elements in the same groups', because of the rigid adherence to the graphic division of the elements to yield a repetition of chemical properties after sixteen atomic weight units.

Newlands, the young English industrialist, arrived at his system, which he named the 'law of octaves', after having published only one table of elements with analogous properties. In his law of octaves of 1865 it is evident that Newlands found a repetition of these properties after seven elements, which has its analogue in the octave of musical theory. In 1864 Odling, who must also be considered as a predecessor in the discovery of the periodic system, devised a system in which a division into main and secondary groups can be recognized. Hinrichs, an American scientist of Danish origin, approached the system from a totally different viewpoint. He was intrigued by the relations which existed between the distances of the planets to the sun, and he took these arithmetical relations as an example for his relations between elements. The classification which he devised in 1867 was a radius system, consisting of a spiral with radii bearing analogous elements. Lothar Meyer, who in 1868 revised the second edition of his book *Die modernen Theorien der Chemie*, was a dyed-in-the wool chemist, which accounts for the fact that his system, created in 1868 but not published until 1871 and then only in a somewhat modified form, can be considered a full-fledged periodic system. In 1869 Mendeleev became the last member of this sextet of discoverers, but because of the depth of his chemical knowledge, he is regarded as the most important discoverer. Also, compared to the other five discoverers, he defended his view for the longest time.

### 3. Shortcomings and Discrepancies in the Periodic System

Although space limitations preclude our following the further development of the periodic system in detail, we must illustrate the deficiencies which the system possessed from its inception as well as the difficulties that appeared later on incorporating some elements or groups of elements into the system. However, we must emphasize the fact that, despite its imperfections, this classification of elements provided yeoman service to chemistry and has continued to do so. We must point out the imperfections to show clearly that this system did not have a theoretical foundation, which sometimes led to confusion. Fortunately, it was at just one of these times that Soddy began to show an interest to the classification of the elements, and he was able to provide the required theoretical background.

First, there was the well-known problem of the elemental pair tellurium-iodine. It had been incontrovertibly proven – despite the fact that some scientists refused to believe it – that the atomic weight of tellurium was greater than that of iodine, whereas it was quite evident from their chemical properties that these elements had to be incorporated in the reverse order, viz., tellurium-iodine, in the periodic system. Second, the pair of elements cobalt-nickel provided an analogous difficulty, but the atomic weight difference was smaller and the differences in chemical properties between these two transition metals were not so pronounced as in the case of tellurium-iodine.

The third pair of elements, argon-potassium, became a problem only after the discovery of the noble gases in the 1890s. The discovery of argon caused a particularly thorny problem as this noble gas with inert character did not fit at all into any of the groups then known. Fortunately, the other noble gases were discovered rapidly after 1894 – the year of the discovery of argon – so that an entirely new group of elements could be incorporated into the system without too much difficulty.

The rare earth metals caused a problem which only became manifest after the initial proposal of the periodic system. Before the 1860s not many of these very similar elements had been discovered, but after the isolation in the 1870s and 1880s of a number of rare earth metals, difficulties arose. This is not surprising inasmuch as these metals constitute a separate family of elements, from its inception the periodic system was one in which elements were arranged according to valence, and the system provided space only for what later would be called main and secondary group elements.

We now know that thorium and uranium are also members of a separate series named the actinides, but these two elements resemble zirconium and tungsten, respectively, in their chemical properties so closely that they did not constitute a real problem. Protactinium was discovered only after the theoretical foundation for the periodic system was known, and therefore its necessary inversion with respect to thorium despite its smaller atomic weight likewise never constituted a problem.

Another difficulty, however, was caused by the discovery of radioactive elements or rather the attempted placement in the periodic system of the disintegration products of the three radioactive series. For the elements discovered at the end of the last century – radium, polonium, and actinium – open spaces had been reserved, but for disintegration products, such as uranium X (1900), thorium X (1902), actinium X (1904–1905), radiothorium and mesothorium (1905), radioactinium (1906), uranium Y (1911), uranium $X_1$ and uranium $X_2$ (1913), at first sight there were no places available. As Soddy contributed to the solution of this problem, we will return to this subject later, but only after having noted that several scientists were badly confused by this discovery of radioactivity, and they made no secret of their doubts about the validity of the periodic system.

## 4. Radioactive Products (Radioelements)

It would be ironic to accuse Henri Becquerel, the discoverer of radioactivity, and Marie Curie, the discoverer of polonium and radium, with bringing the periodic system of the elements into disrepute. Becquerel did nothing more than discover an entirely new phenomenon caused by the spontaneous disintegration of the already known chemical element, uranium. This process, however, gave rise to radioactive disintegration products which could not be directly incorporated into the periodic system, which had already been in existence for almost half a century. Similarly, Marie Curie did nothing more than isolate some of these radioactive disintegration products. Furthermore, space in the periodic system was available for radium and uranium.

The real problems developed only at a later date, when a number of scientists working on the new phenomenon of radioactivity found radioactive disintegration products which did not fit into the periodic system. The scientists were able to determine the atomic weights of these products, but they found the places for these elements already occupied. In particular, scientists such as Theodor Svedberg, Daniel Strömholm, Friedrich O. Giesel, Hans Geiger, Bertram Borden Boltwood, Ernest Rutherford, and Alexander Fleck, who worked on these disintegration products, were virtually in despair, for the periodic system allowed only one element of given atomic weight for each space. The question arose as to what to do with elements such as radium G, actinium C, and thorium D, with atomic weights of 206, 207, and 208, respectively. Were these really elements, simple forms of matter that could not be decomposed? Thus the first decade of the twentieth century was a time of great confusion for chemistry. Elements had been found, which – if they were to be incorporated into the system – would have to share a place in the system with already discovered elements, e.g., the elements named above would have to share a space with lead. What was to be the solution – to create more spaces than the places available in the periodic system? To think of a spatial, i.e., a three-dimensional system? It is therefore not surprising that besides sceptics there were scientists willing to defend the validity of the periodic system with all available means, even by transforming its periods or rows.

One of the most striking discoveries was that the radioactive disintegration product of uranium and radium, the gas which in the course of time received several names and which was finally named radon, not only was one of the radioactive substances but was also found in 1910 to be an element in the newly discovered group of inert gases. Soddy devoted himself to the problems caused by the discovery of radioactive disintegration products.[4] He suggested that the newly discovered radioactive element polonium be considered an isotope of tellurium. Indeed, the concept of isotopy proposed by Soddy helped to solve this complicated problem. In addition to Soddy, however, other chemists made contributions toward solving the problem, e.g., the products radio-lead and radio-tellurium found by Willy Marckwald[5] in bismuth minerals were indentified by André Debierne[6] as the newly discovered element polonium.

## 5. Spatial Systems

Mendeleev believed that a spatial or three-dimensional system was the best way to give each chemical element its correct place. It is true that he did not explicitly set up such a system in 1870, the year of the discovery of the periodic system. However, he recognized the difficulties in incorporating the elements into a two-dimensional system, i.e., on a plane surface. Mendeleev's main difficulties lay in the few rare earth metals then known and in a few dubious elements, such as didymium, for which he sought a cubic system, i.e., one consisting not only of periods and groups of elements but also of rows of elements which formed the third dimension in such a system.

During the years of development of the periodic system several scientists constructed spatial systems. In chronological order we should mention Sir William Crookes' 'pretzel' model (1889) (Figure 1), Antonius Johannes van den Broek's cubic system (1911), Soddy's helical representation (1913), Emil K. Kohlweiler's system (1920), F. H. Loring's wedge-shaped model (1920), John D. Clark's 'arena' model (1921), M. Courtines' spatial system (1925), Charles Janet's cylindrical screw-formed system (1928), a system with which W. Finke wanted to express the modern atomic theory, and D. F. Stedman's conic model (1947). An

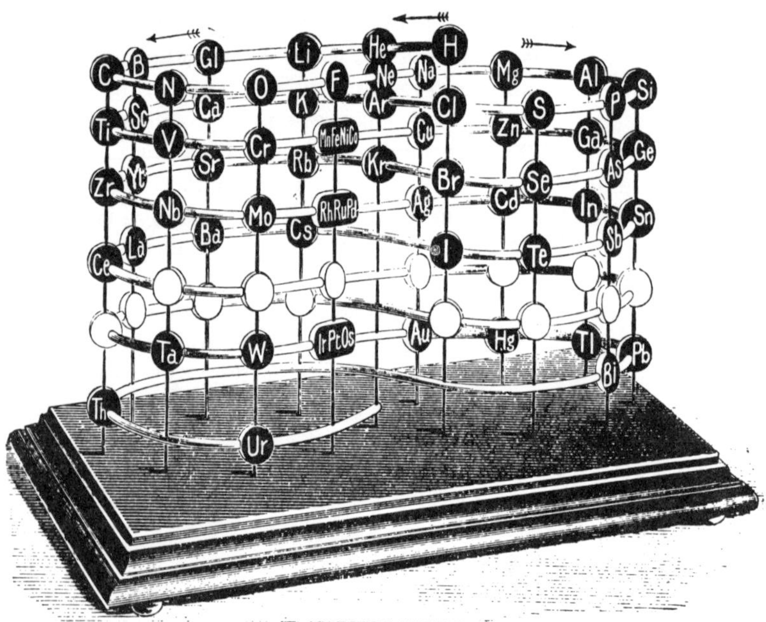


Fig. 1. Crookes' 'Pretzel' Model of the Periodic System. Original copy, made by Crookes himself, on display at the Science Museum, London.

Fig. 2. Antonius van den Broek (1870–1926).

urgent need for such a three-dimensional system arose during the period of difficulties occasioned by the discovery of the radioactive elements, i.e., the period from the discovery of radium to the complete explanation of the existence of radioactive disintegration products.

The first scientist to construct a so-called cubic system was the Dutchman Antonius Johannes van den Broek (1870–1926) (Figure 2)[7], who, at the end of 1912 – before Moseley, Soddy, and Rutherford's work – introduced the hypothesis that the nuclear charge of an atom indicates the place of that element in the periodic system. Although van den Broek did not prove this hypothesis experimentally (he was a lawyer rather than a physicist), it was accepted immediately by physicists.

Van den Broek was interested in finding a system in which all existing chemical elements could be incorporated. In 1907 he published a system based on the $\alpha$-particle. In his research on radioactive radiation, Ernest Rutherford had concluded that $\alpha$-particles were building blocks of matter, but the nature of the $\alpha$-particle was still uncertain at that time. Van den Broek believed that the $\alpha$-particle was half of a helium atom which he named an alphon, with a mass of two atomic weight units. He accepted the arrangement of elements according to increasing atomic weight, but he believed that in all cases each element would differ from its neighbors by two atomic weight units. He thus obtained a periodic system – the alphade system – with space for 120 elements, divided into 8 groups and 15 periods.

In 1911 van den Broek[8] constructed a spatial system intended to express better the periodicity of the properties of the elements. He chose the form of a cube (Figure 3). Accepting Rutherford and Barkla's conclusion that the nuclear charge of an atom must equal half its atomic weight, van den Broek posited, at the end of the year 1912, that the ordinal number of an element in the periodic system is equal to the number of positive charges in the nucleus of its atom. The relationship between atomic number and atomic weight was not as simple as he had first believed. Later research revealed that the relationship between nuclear charge and atomic weight was not a linear one, and in the cases of argon and potassium as well as of tellurium and iodine the positions of the elements had to be reversed. Although van den Broek's periodic system turned out to be incorrect (see, e.g., the splitting of rare earth metals such as $Gd_1$, $Gd_2$, $Gd_3$ and $Dy_1$, $Dy_2$, $Dy_3$), his basic principle of arrangement was correct, for today, of course, elements are arranged in the periodic system in order of increasing nuclear charge.

| | | 0 | I | II | III | IV | V | VI | VII |
|---|---|---|---|---|---|---|---|---|---|
| A | 1 | *He* 4; 4,0 | *Li* 6; 6,9 | *Be* 8; 9,1 | *B* 10; 11,0 | *C* 12; 12,0 | *N* 14; 14,0 | *O* 16; 16 | *F* 18; 19,0 |
| | 2 | *Ne* 20; 20,2 | *Na* 22; 23,0 | *Mg* 24; 24,3 | *Al* 26; 27,1 | *Si* 28; 28,3 | *P* 30; 31,04 | *S* 32; 32,07 | *Cl* 34; 35,47 |
| | 3 | *Ar* 36; 39,9 | *K* 38; 39,1 | *Ca* 40; 40,1 | *Sc* 42; 44,1 | *Ti* 44; 48,1 | *V* 46; 51,06 | *Cr* 48; 52,0 | *Mn* 50; 54,9 |
| B | 1 | *Fe* 52; 55,85 | *Co* 54; 58,97 | *Ni* 56; 58,68 | *Cu* 58; 63,57 | — 60; — | — 62; — | — 64; — | *Zn* 66; 65,37 |
| | 2 | — 68; — | — 70; — | — 72; — | *Ga* 74; 69,9 | *Ge* 76; 72,5 | *As* 78; 74,96 | *Se* 80; 79,2 | *Br* 82; 79,9 |
| | 3 | *Kr* 84; 82,9 | *Rb* 86; 85,45 | *Sr* 88; 87,6 | *Y* 90; 89,0 | *Zr* 92; 90,6 | *Nb* 94; 93,5 | *Mo* 96; 96,0 | *Ru* 98; 101,7 |
| C | 1 | *Rh* 100; 102,9 | *Pd* 102; 106,7 | — 104; — | — 106; — | *Ag* 108; 107,9 | — 110; — | — 112; — | *Cd* 114; 112,4 |
| | 2 | — 116; — | — 118; — | — 120; — | *In* 122; 114,8 | *Sn* 124; 119 | *Sb* 126; 120,2 | *Te* 128; 127,6 | *J* 130; 126,9 |
| | 3 | *Xe* 132; 130,7 | *Cs* 134; 132,8 | *Ba* 136; 137,37 | *La* 138; — | *Ce* 140; — | *Nd* 142; — | *Pr* 144; — | *Sm* 146; — |
| D | 1 | *Eu* 148; — | *Ak C* 150; — | $Gd_1$ 152; — | $Gd_2$ 154; — | $Gd_3$ 156; — | $Tb_1$ 160; — | $Tb_2$ 160; — | *Ak B* 162; — |
| | 2 | $Dy_1$ 164; — | $Dy_2$ 166; — | $Dy_3$ 168; — | *Ho* 170; — | *Er* 172; — | $Tu_1$ 174; — | $Tu_2$ 176; — | $Tu_3$ 178; — |
| | 3 | *Ak Em* 180; — | *Ak X* 182; — | *Yb* 184; 172 | *R Ak* 186; — | *Lu* 188; 174 | *Ta* 190; 181,0 | *W* 192; 184,0 | *Os* 194; 191 |
| E | 1 | *Ir* 196; 193,1 | *Pt* 198; 195,2 | *Au* 200; 197,2 | *Hg* 202; 200,0 | *Th C* 204; 200 | *Tl* 206; 204,0 | *Th B* 208; 204 | *Pb* 210; 207,1 |
| | 2 | $Th_3$ 212; 208 | *Bi* 214; 208 | *Ra F* 216; 210 | $Th_2$ 218; 212 | *Ra C* 220; 215 | $Th_1$ 222; 216 | *Ra A* 224; 219 | *Th Em* 226; 220 |
| | 3 | *Ra Em* 228; 223 | *Th X* 230; 224 | *Ra* 232; 226,4 | *R Th* 234; 228,4 | *Th* 236; 232,4 | *Jo* 238; 230,5 | *U* 240; 238,5 | 242; |

Van den Broek's Cubical System (1911).

Fig. 3. Cubic (Alphade) System, in which the 120 Elements in Groups of Three are Perpendicular to the Design Plane (van den Broek, 1911).

## 6. Theoretical Developments of 1913

The year 1913 was one of the most important in the history of the periodic system. In the same way that the revolutionary year of 1789 can be designated the year of the birth of modern chemistry, since it witnessed the publication of Lavoisier's *Traité de Chimie*, of Higgins' *A Comparative View of Phlogistic and Antiphlogistic Theories*, and of Richter's dissertation *De uso matheseos in chymia*, so could 1913 be designated as the year in which the periodic system of the chemical elements was firmly established on a theoretical basis. In that year Niels Bohr developed his atomic theory, Soddy and others proposed the so-called displacement law and the concept of isotopy, and Henry Moseley introduced the ordinal numbers (atomic numbers) of the elements on the basis of x-ray spectra.

Mendeleev had discovered his periodic system in a manner much like solving a puzzle. He arrived at it by playing a sort of solitaire or patience game with cards on which he had written the chemical data of the chemical elements. It still remains a great discovery, having been discovered at a time in which scientists possessed no concrete data concerning the construction of the elements, let alone the structure of atoms. Therefore it is not surprising that scientists took the atomic weight of the elements as the classifying factor in such a system. The year 1913, however, brought theoretical clarification for the system from a number of different sources.

Bohr clearly showed that the nucleus of the atom consists of particles which define its (positive) charge as well as of other particles (later named neutrons) which, together with the proton, determine the mass of the atom. He also proposed that electrons, revolving in their orbits around the nucleus of the atom, determine the chemical properties of the element. Strictly speaking, Moseley confirmed experimentally on the basis of x-ray data what van den Broek had already stated intuitively, viz., that the charged nuclear particles indicate the order of the elements in the periodic system. He was able to substitute as ordinal numbers of the elements in the periodic system the so-called atomic numbers in place of the atomic weights used up to that time.

Independently of each other, Soddy and Kasimir Fajans demonstrated, with their so-called displacement law and the fact that the $\alpha$-particle consists of two protons and two neutrons, that in radioactive disintegration the atomic number of an element decreases by 2 and the atomic mass decreases by 4 units. After the introduction of Soddy's isotopy concept it became clear that what van den Broek thought valid for all elements – a difference in atomic weight of half of an $\alpha$-particle – could indeed be valid for isotopes of different elements. This law was thus a confirmation of previous discoveries, and it brought to an end many uncertainties in assigning spaces to radioactive elements and other elements in the radioactive series. As early as 1911 Soddy foresaw the displacement law, at least the part concerning the emanation of $\alpha$-particles: 'the product derived by the loss of an $\alpha$-particle is . . . in each case the step being from the family of even valency into the next, the family of odd valency being missed'.[9] Soddy reiterated this conclusion in his article of 1913, in which he developed the full displacement law and pointed out the related work of his co-worker Alexander Fleck and that of A. S. Russell.[10] It is unfortunate that Soddy did not illustrate his system with a more convincing scheme. The printed scheme on page 3 of his book of 1913,[11] which is a reproduction from a paper that he read in

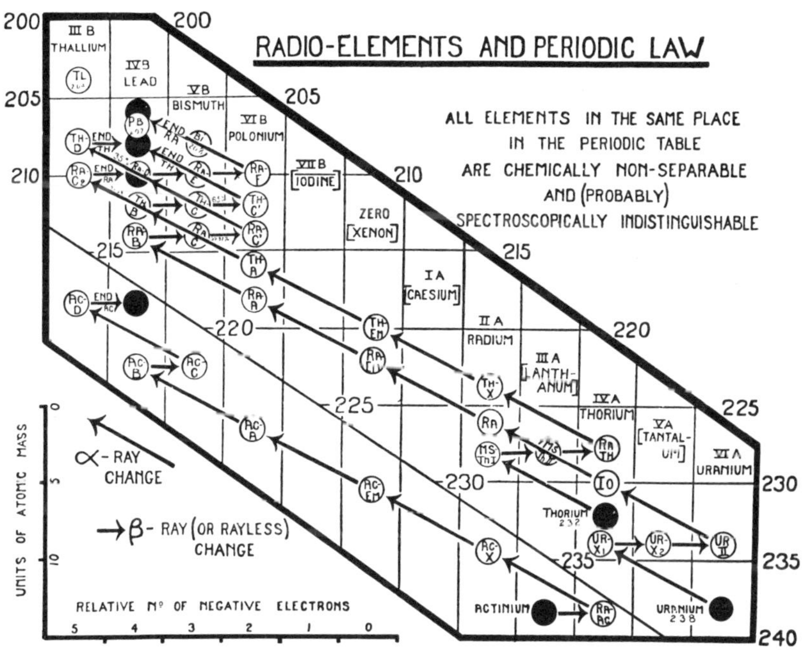


Fig. 4. Scheme Demonstrating the 'Displacement Law'.

1913 before Section B of the British Association (Figure 4), is divided into differences in atomic mass of 5 units, while a division into differences of 4 units would have illustrated the displacement law more convincingly.

## 7. Soddy and the Periodic System

Although during the years around 1913 van den Broek played an important role in the development of the atomic model, in the period of the development of atomic physics, as a nonexperimentalist, his role was less influential than that of the practical experimental physicists of those years, of whom we may name only Rutherford, Fajans, Fleck, Russell, and Soddy. In common with van den Broek, Soddy was interested in the incorporation into the periodic system of the existing chemical elements and the radioactive disintegration products, and this interest dates from 1904, as evidenced, e.g., by his regular broaching of this subject in the section that he wrote for the *Annual Progress Reports to the Chemical Society*. Soddy also devised a spatial system of his own in which he tried to include the elements and radioactive particles, which he published two years later than van den Broek did. Although Soddy had intended to publish it in 1911 as van den Broek had actually done, he gave concrete form to his intention only in 1913.

In the chapter on 'Classification and Nomenclature of Radio-Elements' in Part 1 of his 1911 book *The Chemistry of the Radio-Elements*, Soddy had already involved the periodic system in his conclusions which foreshadow his displacement law. His ideas became more definitive in the second part of his book,[11] which he wrote in 1913 and which bears the subtitle *The Radio-Elements and the Periodic Law*. From page 9 on, Soddy extensively defended his choice of a three-dimensional system (Figure 5), influenced by the so-called 'pretzel' model, which had been proposed in 1888 by Sir William Crookes (Figure 1).[12]

Soddy's own variation of this system is characterized by the omission of any head or tail. The gradual transition in properties of the elements within their groups and the connective properties of the noble gases, which cannot be expressed in two-dimensional systems, can be shown in such three-dimensional systems. In order to avoid distorting history we must note that neither Crookes nor Soddy were the first scientists to

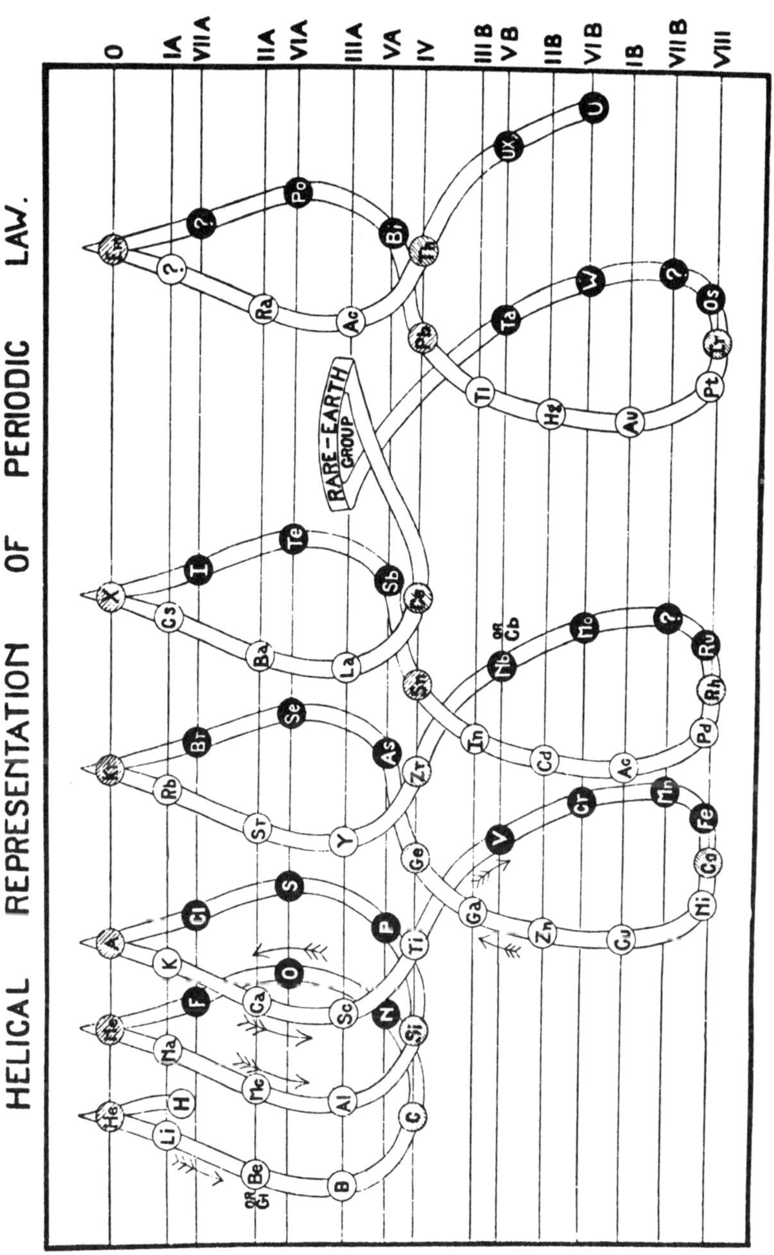


Fig. 5. Soddy's Three-Dimensional Periodic System.

realize this. At least seven years before Mendeleev first discovered the periodic system, the French mineralogist and geologist Béguyer de Chancourtois, in 1862, proposed his cylindrical model construed as a continuous series of elements. Here too, the groups of the alkali metals and the halogens were contiguous with each other rather than extremities. Of course, since the noble gases had not yet been discovered, they could not form a link between the alkali metals and the halogens. Because of technical printing considerations, Soddy represented his system, which he called a 'helical representation of the periodic law', in a plane. However, he stated explicitly that some elements (lettered black upon a white background) are supposed to lie *above* the plane, while others (lettered white upon a black bakground) lie *below* it. The noble gases, along with carbon, silicon, titanium, etc. (the elements of the fourth group) as well as cobalt, rhodium, and iridium are the only elements *in* the plane.

As we have seen, Soddy had alrady shown that the foundations of the system were totally clear to him. The principal advantage of his helical system over the plane periodic table (Figure 6) as given by Soddy on Page 10 of Part II of his book is, as he himself suggested, the elimination, by making *two* different kinds of 'screw threads', of the artificial open places arising in the groups 0 to VIII. This simultaneously involved a natural division of the elements into those of the so-called main groups and those of the secondary groups. Strictly speaking, only the main groups possess a group 0 (the noble gases) and the secondary ones a group VIII (the transition metals). Furthermore, it was perfectly clear to Soddy that any rigid division of the elements into A periods and B periods does not hold in his helical system, for he remarked that the transition from elements of the A series to those of the B series occurs in the middle of the series, i.e., in Group IV. Thus antimony and tellurium, for example, had to be considered as belonging to the A series rather than the B series, a correct assignment which could be proven only by knowing the electronic configuration of the atoms of the elements. It was only after 1921 that scientists were able to define main group elements as those possessing structures in which the last electron had an *s* or a *p* electronic configuration and subgroup elements as those having a *d* electronic configuration. Consequently, today the main group is assumed to continue with the elements gallium-indium-thallium. Soddy's suggestion was thus a very positive contribution to the periodic system.

# PERIODIC TABLE OF THE ELEMENTS

| | Group O. | Group I. | Group II. | Group III. | Group IV. | Group V. | Group VI. | Group VII. | Group VIII. | | |
|---|---|---|---|---|---|---|---|---|---|---|---|
| Hydrogen 1.008 | Helium He 3.99 | Lithium Li 6.94 | Beryllium Be 9.1 | Boron B 11.0 | Carbon C 12.00 | Nitrogen N 14.01 | Oxygen O 16.00 | Fluorine F 19.0 | | | |
| | Neon Ne 20.2 | Sodium Na 23.00 | Magnesium Mg 24.32 | Aluminium Al 27.1 | Silicon Si 28.3 | Phosphorus P 31.04 | Sulphur S 32.07 | Chlorine Cl 35.46 | | | |
| A | Argon A 39.88 | Potassium K 39.10 | Calcium Ca 40.07 | Scandium Sc 44.1 | Titanium Ti 48.1 | { Vanadium V 51.0 | Chromium Cr 52.0 | Manganese Mn 54.93 | Iron Fe 55.84 | Cobalt Co 58.97 | Nickel Ni 58.68 |
| B | | Copper Cu 63.57 | Zinc Zn 65.37 | Gallium Ga 69.9 | Germanium Ge 72.5 } | Arsenic As 74.96 | Selenium Se 79.2 | Bromine Br 79.92 | | | |
| A | Krypton Kr 82.92 | Rubidium Rb 85.45 | Strontium Sr 87.63 | Yttrium Yt 89.0 | Zirconium Zr 90.6 | { Niobium Nb 93.5 | Molybdenum Mo 96.0 | — | Ruthenium Ru 101.7 | Rhodium Rh 102.9 | Palladium Pd 106.7 |
| B | | Silver Ag 107.88 | Cadmium Cd 112.40 | Indium In 114.8 | Tin Sn 119.0 } | Antimony Sb 120.2 | Tellurium Te 127.5 | Iodine I 126.92 | | | |
| A | Xenon Xe 130.2 | Caesium Cs 132.81 | Barium Ba 137.37 | | [Lanthanum La 139.0 | Cerium Ce 140.25 | Praesodymium Pr 140.6 | Neodymium Nd 144.3 | | Samarium Sa 150.4 | |
| | Europium Eu 152.0 | | Gadolinium Gd 157.3 | | Terbium Tb 159.2 | Dysprosium Dy 162.5 | | Erbium Er 167.7 | | | |
| | | Thulium Tm 168.5 | Ytterbium Yb 172.0 | Lutecium Lu 174.0] | | { Tantalum Ta 181.5 | Tungsten W 184.0 | — | Osmium Os 190.9 | Iridium Ir 193.1 | Platinum Pt 195.2 |
| B | | Gold Au 197.2 | Mercury Hg 200.6 | Thallium Tl 204.0 | Lead Pb 207.10 } | Bismuth Bi 208.0 | (Polonium) | — | | | |
| A | Radium Emanation 222. | — | Radium Ra 226.0 | Actinium | Thorium Th 232.4 | Uranium $X_2$ (Brevium) | Uranium U 238.5 | | | | |

Fig. 6. Soddy's Periodic System.

In critically considering Soddy's periodic system, we must mention the place of the rare earth elements. These form a continuous group positioned on a 'graceful little bow' between the elements cerium and tantalum. Unfortunately, here he missed a chance to properly position these elements, of which more was known in those years than in the entire period of discovery of the system. The rare earths were properly placed by Thomas Bayley, Julius Thomsen, and Alfred Werner.

In 1922 Carl Renz[13] used Soddy's helical system to incorporate the rare earth elements on a subspiral between two helices (Figure 7). In comparing Renz's contribution with that of Soddy, we find that Soddy had actually suppressed element 72. He should have reserved a place before tantalum for a homologue of zirconium (Hafnium was not discovered until 1923). It is also unfortunate that Soddy[4] in 1911 considered thorium as a homologue of cerium, thus forming a second series of rare earth metals, but that this did not work out in his system. In his system Soddy placed a number of question marks at several places for

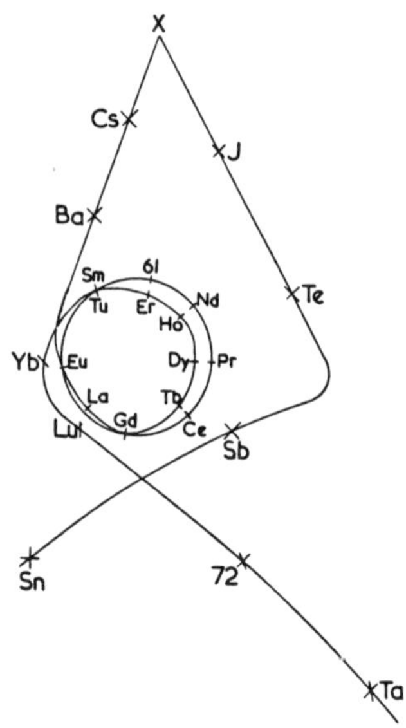


Fig. 7. Incorporation of the Rare Earth Elements in Soddy's 'Pretzel' System (Carl Renz, 1922).

elements which had not yet been discovered (technetium, rhenium, astatine, and francium), a quite acceptable procedure, since specific places had to be reserved for elements still to be discovered.

In Soddy's helical system, among the radioactive disintegration products we see only the indication $UX_2$ at the place between thorium and uranium. In his accompanying periodic table (Figure 6), Soddy added between brackets the name brevium for this product. In his book[11] he referred to this name, which he regarded as very suitable for the short-lived isotope designated uranium $X_2$ by its discoverers Fajans and Göhring[14] in 1913. It is actually a short-lived isotope of element 91, protactinium, which had not yet been discovered. Considering the complete radioactive series, not less than 34 elementary substances, named by Soddy as elements, were known in 1913, but there were available only ten free places in the periodic system. Five of these ten places belonged to elements which were already well known, such as lead, bismuth, and uranium. The other places belonged to elements discovered as a result of their radioactivity. Thus radium, radon, polonium, and actinium were new elements, but the places 87 and 89 were still reserved by Soddy for elements yet to be discovered. Indeed, among the disintegration products of the radioactive series, no isotopes of elements were found which could claim one of these free spaces.

For element 91, the place that Soddy designated brevium, Otto Hahn and Lise Meitner in 1917 discovered the long-lived isotope which they named proto-actinium. At about the same time Soddy also independently discovered this element, as eka-tantalum, in collaboration with John A. Cranston (Chairman of the Glasgow Section of the Society of Chemical Industries) in pitchblende residues from India and Joachimsthal and thus replaced brevium by protactinium.[15]

With his periodic system Soddy wanted to express clearly that an end had come to the time of chaos initially caused by the discoveries of the radioactive elements. But this is not Soddy's only contribution. At the same time he rehabilitated the periodic system, which was in danger of being discredited. He also used this system to predict new details in the properties of expected new elements, which would then have to be confirmed experimentally. Thus just as did the famous discoverer Mendeleev, Soddy also treated chemistry as an inductive science, and he assigned the periodic system a central role within it. This is the reason for Soddy's predilection for the title 'periodic law' for this natural system, table, law, or simple classification of chemical elements.

*Veenendaalkade 463*
*2547 Den Haag*
*The Netherlands*

## Notes

[1] van Spronsen, J. W.: 1969, *The Periodic System of Chemical Elements: A History of the First Hundred Years*, Amsterdam: Elsevier Publ. Co., p. 9.
[2] Cassebaum, Heinz and Kauffman, George B.: September, 1971, 'The Periodic System of the Chemical Elements: The Search for Its Discoverer', *Isis* **62**: 3, 313–327.
[3] van Spronsen, *op. cit.*, Chapter 5.
[4] Soddy, F.: 1911, 'The Chemistry of Mesothorium', *J. Chem. Soc.* **99**, 72–83.
[5] Marckwald, W.: 1902, 'Ueber den radioactiven Bestandtheil des Wismuths aus Joachimsthaler Pechblende', *Ber.* **35**, 4239–4241; **36**: (1903): 2662–2667.
[6] Debierne, A.: 1904, 'Sur le plomb radioactif, le radio-tellure et le polonium', *Compt. Rend.* **139**, 281–283.
[7] van Spronsen, J. W.: 1976, 'The Classification of Elementary Particles', *Chem. Weekbl., Mag.* **72**, 634–635.
[8] van den Broek, A.: 1911, 'Das mendelejeff'sche "kubische" periodische System der Elemente und die Einordnung der Radio-elemente in diesem System', *Physik. Z.* **12**, 490–497.
[9] Soddy, F.: 1911, *The Chemistry of the Radio-Elements*, London: Longmans, Green & Co., Part I, p. 30.
[10] Soddy, F.: 1913, 'The Radio-Elements and the Periodic Law', *Chem. News* **107**, 97–99.
[11] Soddy, F.: 1914, *The Chemistry of the Radio-Elements: Part II: The Radio-Elements and the Periodic Law*, London: Longmans, Green & Co. Also see *Annual Progress Report to the Chemical Society for 1913* **10**: (1914): 262–288.
[12] Crookes, W.: 1888, 'Elements and Meta-Elements', *J. Chem. Soc.* **53**, 487–505.
[13] Renz, Carl: 1922, 'Die seltenen Erden im periodischen System', *Z. anorg. allgem. Chem.* **122**, 135–145.
[14] Fajans, K. and Göhring, O. H.: 1913, 'Über das Uran $X_2$ – das neue Element der Uranreihe', *Physik. Z.* **14**, 877–884; 'Über die komplexe Natur des Ur X, *Naturwissenschaften* **1**: (1913): 339.
[15] Soddy, F. and Cranston, J. A.: 1918, 'The Parent of Actinium', *Nature* **100**, 498–499. 'The Parent of Actinium', *Proc. Roy. Soc. (London)* **94A**: (June 1, 1918): 384–405.

PART IV

# RECEPTION OF SODDY'S WORKS AND IDEAS

A. N. KRIVOMAZOV

# 6. THE RECEPTION OF SODDY'S WORKS IN THE U.S.S.R.

*Soddy's books, originally published in England, have been translated and published more frequently in Russia than in any other country, and their success in Russia has been both great and consistently maintained up to the present day. Soddy's ideas were popularized by such eminent Russian scientists and writers as V. I. Vernadsky, L. A. Chugaev, N. A. Shilov, N. A. Morozov, D. D. Khmyrov, V. A. Borodovsky, V. G. Khlopin, and Maxim Gor'kiĭ.*

## 1. Soddy's Books in Russian Translation

It is of considerable interest to the historian of science in general and to the historian of twentieth-century physics in particular that in no other country of the world have Soddy's books, which were originally published in England, been translated and published as frequently as in Russia. Indeed, as will be shown below, the success of his works in Russia has been both very great and consistently sustained throughout the years.

How can one explain why Russian readers have been interested precisely in Soddy's books and not in those of other research pioneers in radioactive phenomena, e.g., M. Curie, E. Rutherford, S. Meyer, E. von Schweidler, and others? As we see it, the reason for Soddy's books being so popular in Russia lies in the specific attitude of Russian scientists not only towards the results of the study of radioactive phenomena but also towards the theoretical interpretation of these results. In order to clarify this reason, we must remind the reader of the most urgent problems of the chemical atomic theory which are nearly a hundred years old.

In the world scientific literature of the second half of the nineteenth century one can find broad discussions of the structure of matter and its relation to energy [1–3]. Ideas about the complex structure of atoms of chemical elements representing various modifications of Prout's hypothesis [4] as well as ideas about the dissociation of elements and the

*George B. Kauffman (ed.), Frederick Soddy (1877–1956)*, 115–140.

experiments devoted to dissociation of elements under certain conditions (e.g., high temperatures and pressures) [1, 2, 5] have received considerable attention. At the end of the nineteenth and beginning of the twentieth centuries these ideas were also developed by many Russian scientists, including A. T. Bazarov, N. N. Beketov, N. A. Umov, F. M. Flavitsky, T. O. Yarkovsky, and others. For a long time these ideas were of an abstract, speculative nature and, while being sharply rejected by some researchers, enjoyed warm support from others.

At the turn of the century a revolution in physics took place which was associated with the discovery of x-rays, radioactivity, and the electron. With these discoveries ideas about the complex structure of atoms have obtained an important experimental confirmation. An essential factor that prevented the acceptance of new discoveries was Wilhelm Ostwald's criticism of concepts of the atomic theory, a criticism based on energetics. Russian scientists, however, who belonged to a country with well-developed materialist traditions, demonstrated a sharply negative attitude towards energetics. As is well known, D. I. Mendeleev [6] (p. 98), A. G. Stoletov, N. N. Beketov, D. P. Konovalov, N. A. Umov, I. A. Kablukov, N. S. Kurnakov, and other Russian physicists and chemists defended the atomic-molecular theory and criticized energetics [7] (p. 227). It should be noted that during the last years of his life Mendeleev adopted an extremely negative position towards the ideas of dissociation of the elements as well as towards the theory of radioactive decay which was established by the articles written by Rutherford together with Soddy [6] (p. 94).

Since there was no large volume of the original basic research on radioactivity in Russia [8] (p. 15), foreign publications on this topic stirred up considerable interest. From 1901 to 1914 in the *Journal of the Russian Physical and Chemical Society* one could find comprehensive reviews of foreign as well as of Russian papers on radioactivity written by Yu. S. Zal'kind [8] (pp. 37, 210). Furthermore, in Russian scientific and popular-science journals there were translations of articles by well-known foreign scientists who studied radioactive phenomena. Apparently, the first publication of Soddy's to appear in Russian was a 1903 translation in the journal *Fizicheskoe obozrenie* (*Physical Review*) of Ramsay and Soddy's paper 'Gases Occluded by Radium Bromide' [9].

The journal translations, however, could not satisfy the interest of the broad masses of intellectuals – the teachers, students, engineers, and self-educated persons – in this new sphere of science. At the beginning of the twentieth century in Russia there was an urgent need for books about radioactivity that were suitable for broad circles of readers. In Moscow and St. Petersburg in 1903 three translated books on radioactive phenomena, written by F. Giesel [10], K. A. Hofmann [11], and F. Soddy [12], were published.

Compared to the two other books, Soddy's book was especially interesting for Russian physicists and chemists since its author, who was a co-worker of Rutherford's, had developed the foundations of the theory of radioactive decay, and during his joint work with Ramsay he had obtained additional experimental proofs of the validity of this theory. Other valuable qualities of Soddy's book were its visual clarity, definitiveness, and beautiful popular style. According to our point of view, however, the true reason for the attractiveness of this and other books of Soddy's for Russian readers was the broad philosophical coverage of various problems of the interrelationships between matter and energy, problems which are of the utmost interest to Russian scientists.

As is well known, in the autumn of 1903 Soddy began giving lectures on radioactivity at University College, London. These lectures were intended for a broad audience which was consistently informed by Soddy concerning the most important works in this field. As the lectures were being given, they were published in the weekly issues of the English journal *The Electrician*. The efficiency of the Russian publishers can be illustrated by the fact that as soon as this journal finished publishing Soddy's lectures, their translation appeared in St. Petersburg in the form of a book as *Radioactive Phenomena* [12]. Although the title page of this Russian edition contains the name of neither its translator nor editor, its reverse side presents a valuable historic reference: 'Printed by order of the Imperial Russian Technical Society'. One can assume that both the translation and editing of this book were done through the efforts of the members of this technical society.

In 1905 a new book of Soddy's was published in Russia – *Radioactivity. An Elementary Treatise from the Standpoint of the Disintegration Theory* [13]. The preface, in the form of a translator's comment, was written by its translator, F. N. Indrikson, who had studied radioactivity under I. I. Borgman's supervision in the Physical Institute of St. Petersburg University [8] (pp. 19–20).

In his note of November 2, 1904, in which he compared books about the passage of electricity through gases by J. J. Thomson, *Radioactivity* by E. Rutherford, and *Radioactivity* by F. Soddy, Indrikson remarked that Soddy's book is written for a broader audience and that 'its important advantage is the presentation of radioactive phenomena from the point of view of the atomic disintegration theory created by Rutherford and the book's author' [13].

The greatest success in Russia, however, belonged to Soddy's famous scientific best seller, *Interpretation of Radium*, which was published in Russian under the title *Radium and its Clue*. This book was published six times in Russia in different forms [14–19]. The translator of the first Moscow edition of this book, N. A. Shilov, Professor at the Moscow Technical School, gave a very high evaluation in his preface to Soddy's scientific works as well as to his activity as a popularizer of science.

Shilov wrote about Rutherford and Soddy's works:

> The names of Rutherford and Soddy head the papers in which the theory of atomic decay was formulated for the first time; over a brief period of time this theory has accumulated a great number of brilliant confirmations and predictions. The theory of atomic decay became a necessary working hypothesis not only for the special processes of radioactive phenomena but also for the entire broad new field of the chemistry of atoms, which is being explored successfully through this theory [14] (p. ix).

According to Shilov, Soddy's participation in the development of the theory of radioactive decay ensures that the book translated by him is characterized by a high degree of scientific authenticity: 'The participation of Soddy in the development of the basic concepts and theories of radioactivity makes his ideas especially authoritative and historically important: they carry the stamp of the primary source' [14] (p. x).

In his description of Soddy's book Shilov mentions Soddy's comprehensive and competent review of the sequence of scientific discoveries that resulted in a theoretical interpretation of the nature of radioactive phenomena. At the same time, says Shilov, 'Soddy's book is not just a compilation; it gives an idea of the gradual, stubborn scientific labor which was necessary for the study of radioactivity and which by no means always led immediately to the desired goal' [14] (p. x).

Shilov's analysis of the stylistic features of Soddy's book is of great interest:

> Soddy's work can be singled out from the numerous books written about radioactivity by the nature of its presentation. The author always remains on a rigorously scientific plane when he makes use of experimental data, and he presents the most essential material in a

popular form, omitting only the details. But at the same time, in his presentation he can employ not only uncommon simplicity, clarity, and briskness but also the breadth of a philosophical approach and a touch of artistic taste. Even when the author is carried away by his imagination, which is quite noticeable at times, one can sense the spontaneity and frankness of the original creative thought [14] (p. x).

According to Shilov, this book would primarily be useful to those who want to become acquainted for the first time with the new field of research and to arrive at a clear understanding of its achievements and tasks:

Soddy's book presents the nonspecialist reader with a picture of the unknown up to now, with mysteries of nature that are truly exciting and thrilling, and it uses vivid description, sometimes broad images, to speak about what has been accomplished in this field and what remains to be accomplished by future generations. The author brings the question out of the realm of purely scientific problems to the level of a deep philosophical issue, and he goes into a beautiful flight of scientific thought [14] (p. xi).

As Shilov believed, all these qualities might bring success to this rigorously scientific yet highly emotional book of Soddy's: 'Much has been said about radium, but hardly anyone has been able to say his word with such a belief in science, such a deep conviction in a broad interest of the subject presented, and to put it into such beautiful and exciting words [14] (p. x).

In 1910 an Odessa edition of Soddy's book was published simultaneously with the Moscow edition. The editor of the Odessa edition of *Radium and its Clue* was D. D. Khmyrov, a laboratory assistant at Novorossiisk University. Under his editorship, the translation of *Radium and its Clue* was published three times by the Odessa publishing house 'Mathesis' [15–17]. To his third edition (1923), which represented a translation from the fourth revised and supplemented English edition, Khmyrov wrote an introductory note, 'From the Translation's Editor', in which he briefly evaluated the significance of Soddy's book in the modern popular science literature. In Khmyrov's words,

In the popular science literature on radium Professor Soddy's book occupies a place which is as exclusive as that of the scientific works of its author – one of the founders of the theory of atomic decay – in the specialized literature. Inspiration and enthusiasm flow from each page of the book, and therefore the text equally thrills outsiders and specialists in the field; on the other hand, the broad generalizations, the powerful flight of the author's scientific imagination provoke an intense interest in the very facts which may sometimes he cumbersome and complex but which are presented with wonderful clarity and sharpness.

Khmyrov's correspondence with Soddy mentioned in this note and the cited fragment of Soddy's letter are of considerable interest. In Khmyrov's words:

The present translation has been made from the English edition of 1922, which represents a reprint of the fourth edition published in 1920. In this edition the author supplemented the book with a second part devoted to the theory of the structure of the atom. When beginning the translation, we thought it appropriate to ask for the author's permission and to inquire whether he would like to make certain changes or additions. We obtained the following reply (of April 6, 1923): 'As far as I am concerned, I would be completely satisfied if you translate the new and expanded edition into Russian.... The new edition is still in agreement with present developments, and I do not think that any changes are required'.

'According to this recommendation of the author', wrote Khmyrov, 'we have made a complete translation of his book, without any esssential changes' [17].

Khmyrov's new edition of Soddy's book *Radium and the Structure of the Atom* appeared in 1923, and in 1924 the Moscow [18] and Leningrad [19] editions of this book were published. It is interesting to note that this book of Soddy's was published five times in England, once in Germany, twice in France, and six times in Russia.

Translations of two other books by Soddy, *The Chemistry of the Radioelements* [20] and *Matter and Energy* [21, 22], appeared in Russia nearly at the same time (1913), as the book *Matter and Energy* was being published simultaneously by two Moscow publishers in different translations.

V. A. Borodovsky, one of the first Russian radiologists, worked from April, 1908 to January, 1910 on the course of radioactive phenomena in the laboratories of J. J. Thomson, E. Rutherford, and others [8] (pp. 83–88). In 1913 the St. Petersburg publishing house 'Education' published Soddy's book *The Chemistry of the Radioelements* [20], which was translated by R. Shmidt and edited by Borodovsky. In his preface to the Russian edition Borodovsky wrote:

Soddy's book *The Chemistry of the Radioelements* represents an exceptional phenomenon in the literature on radioactive substances.... Separate remarks of various authors concerning the chemical properties of the radioelements were scattered over numerous specialized journals. Soddy undertook the task of critically analyzing the research on the chemistry of radioelements that had been known up to that time, and in his book he offers a systematic review of the chemical properties of each radioelement separately. The simultaneous consideration [by Soddy] of the physical and chemical properties of a radioelement completely establishes its proper nature in each individual case.

It is interesting that Borodovsky mentioned the rapid progress in the field of research on radioactive phenomena (the discovery of radioactive 'branches' in the radioactive series of uranium and thorium, etc.) achieved during the last two years: 'In spite of the fact that Soddy's book has been published in England so recently (1911), a great amount of data which pertained mainly to physical constants had to be replaced by the results of the most modern studies' [20].

In 1913 the Moscow publishing house 'Nature' published Soddy's book *Matter and Energy* [21], translated by S. G. Zaimovsky and edited by N. A. Morozov. Morozov (1856–1946) was a Russian revolutionary (In 1883–1905 he was imprisoned in the Schlisselburg fortress in connection with the *Narodnaya Volya* (revolutionary-terrorist organization) case), an author of numerous works on chemistry, physics, astronomy, mathematics, and history, and an honorary member of the U.S.S.R. Academy of Sciences since 1932. During the second decade of the twentieth century his book *The Periodic Systems of the Structure of Matter* enjoyed great popularity. In Soddy's book Morozov inserted his own version of the periodic system of the chemical elements (whose shortcoming was the absence of a place for hydrogen), wrote 'Preface to the Russian Translation', and annotated Soddy's text in a number of places. In his preface Morozov introduced Soddy and his book to the Russian reader in the following words:

> Frederick Soddy is the third and youngest member of the brilliant trio of contemporary British scientists in the field of radioactive substances.
>
> His scientific activity started as an assistant of Rutherford's and later of Ramsay's; he participated in their two most important discoveries in this field.
>
> Ramsay's discovery of helium [during the decay of radium emanation] was made together with Soddy. The theory of disintegration of the atoms of the periodic system of elements into helium was formulated by Rutherford together with Soddy.
>
> Now [1913] he is a lecturer on physical chemistry and radioactivity at Glasgow University in Scotland. His most important previous works, *Radium*, *The Chemistry of the Radio-elements*, *Popular Lectures on the Nature of Radium* are already available in Russian translation. His present book [*Matter and Energy*] represents, as it were, a philosophical generalization of the concepts concerning the structure of matter and energy that is being produced by the latest trends in physical science [21] (p. 11).

We do not know of any other editions of Soddy's books in Russian. Apparently, they do not exist. We shall now trace the response of Russian scientists and other readers to Soddy's books to the problems that he discusses.

## 2. Scientific Discovery and its Trustworthiness: F. Soddy, G. N. Antonoff, and the Discovery of Uranium-Y

When Soddy investigated the first disintegration products of the uranium series in 1909 [23–29], he attempted to find a new radioelement in solutions of uranium-X. He believed that complex $\beta$-radiation (hard and soft) of uranium-X might indicate the existence of an undiscovered radioelement. However, he encountered such great experimental difficulties, and the theoretical interpretation of his own results seemed so dubious that he gave up this very interesting and difficult experimental work.

In 1911 Georgii Nikolaievich Antonoff, an unknown 31-year-old Russian research student of Ernest Rutherford's, published the results of his experimental research on the strange behavior of uranium-X in the *Philosophical Magazine* [30]. Antonoff described his chemical method in detail and claimed the discovery of a new radioelement which he called uranium-Y. In a private letter dated May 16, 1911 Rutherford wrote to his American colleague and friend Bertram Borden Boltwood about this discovery:

> Antonoff, who has been working like a slave for the last year, believes he has found a new product emitting $\alpha$-rays [an error; this should read $\beta$-rays] of period about 1.5 days in purified uranium. He has done a great deal of work, and it looks as if it may be correct [31].

Soddy had twice discussed this result of Antonoff's in 1911 – in his valuable book *The Chemistry of the Radio-Elements* [20] (p. 60) and in his 1911 *Annual Progress Report on Radioactivity* [54] (pp. 274–5). Soddy noted that it was desirable to verify Antonoff's discovery of uranium-Y.

In 1913 Soddy's demonstrator Alexander Fleck was unable to confirm Antonoff's result which he discussed in his articles of March and May of 1913 [32, 33]. Antonoff received these pubications of Fleck's comparatively calmly. In his Russian book *Uranium Disintegration Products*, he mentioned that the separation of uranium-Y from uranium-X 'offers almost unsurpassed experimental difficulties.... It was only after an entire year of hard work that I managed to find conditions under which this experiment could be easily repeated in my own hands' [34].

In the August, 1913 issue of the *Philosophical Magazine* there was published Antonoff's letter to the editor, in which he assumed that

Fleck was unable to separate uranium-Y only because of great experimental difficulties [35]. Soviet historians of science L. L. Zaitseva and N. A. Figurovsky noted that this controversy

might be prolonged, because Antonoff could not purify uranium from uranium-X, and Soddy had not described in detail his methods of uranium purification. In his letter to Rutherford, Antonoff asked him to solve the problem. For this purpose Antonoff was ready to leave for England to repeat his experiment in the presence of experts. On the suggestion of Rutherford, who believed that the best way to solve the controversy was for Antonoff to repeat his experiments with the uranium sample purified by Soddy himself, Soddy had sent to Antonoff 60 grams of purified uranium nitrate, from which Antonoff separated uranium-Y in the same manner as from his own uranium preparation [8] (p. 114).

On July 3, 1913 Soddy wrote to Rutherford:

As regards Antonoff, I should be delighted if I can be of any service in enabling him to repeat his Ur-Y experiments and will gladly send him some of my purified uranium nitrate. Tell him to let me know when he wants it and how much he would like [36].

After reproducing experiments with Soddy's uranium nitrate, Antonoff wrote to Rutherford on August 16, 1913:

Dear Professor Rutherford,

I am very glad to tell you that I found Ur-Y in the uranium loaned to me by Soddy. I have obtained even much better curves than previously, showing the presence of the product with a period of 1.2–1.5 days in uranium in small quantity compared with Ur-X.

Now I would like to know your opinion as to what to do next?

First of all, must I write to Soddy about it or must he get to know about it from the periodicals? My personal view is that it is desirable to send a very short letter to the Editor of the *Phil. Mag.* simply stating that I have separated Ur-Y from the preparation loaned to me by Soddy, which according to him is not supposed to contain Ur-Y.

Simultaneously I am thinking of sending a longer paper to *Phys. Zeit.* or *Le Radium*, in which I intend to describe properly my method of separation of Ur-Y, and to express definitely my belief that Ur-Y is a branch product.

I hope that after this Soddy will be convinced. If not, I repeat, again, I am quite prepared to perform the experiment in the presence of witnesses at any time and wherever you like [37].

At a meeting of the Chemistry Section of the Russian Physical-Chemical Society on September 12, 1913 Antonoff said of his success in isolating uranium-Y from Soddy's preparation:

Thus I managed to prove the existence of uranium-Y and to show that it is generated by uranium. In its disintegration process, therefore, uranium is transformed simultaneously into two products: uranium-X (predominately) and uranium-Y (in minor amounts, because it must be considered as a branch product) [38] (p. 1464).

In the Rutherford-Soddy correspondence one may find Soddy's letters to Rutherford of October 2, 23, and 28, 1913, in which are discussed some details of the discovery of uranium-Y. The most interesting is the first letter (October 2) containing Soddy's impressions of Antonoff's recognition of his discovery with Soddy's uranium nitrate. In this letter Soddy wrote:

> Antonoff wrote me a few days ago that he has confirmed his results with my uranium as well as with his own, and I am now trying to repeat his work myself in conjunction with Fleck.... My own opinion is that he is probably correct on the general conclusion but that his account in the *Phil. Mag.* was so imperfect that anyone trying to repeat his experiment would be certain to fail. For this reason, as I told him, I wish he would publish full details of his actual experiments without delay. There was certainly an inference in Fleck's paper that Antonoff's results might be due to thorium, but he only stated that he himself had only been able to obtain similar effects when thorium was present. He did not conclude that Antonoff's results were due to thorium as Antonoff states in his letter that you sent me [39].

A short note by Antonoff was published in the December, 1913 issue of the *Philosophical Magazine*, in which he confirmed the isolation of uranium-Y from Soddy's preparation and also mentioned a comparative activity of uranium-Y (about 2% of the activity of uranium-X). Antonoff believed that uranium-Y was the parent of actinium: 'the actinium series branched off at this point' [40].

In December, 1913 Soddy took up the study of the problem of the existence of uranium-Y. Of course, he was interested in preserving the prestige of his laboratory. If Antonoff were to prove the existence of errors in Fleck's results, that might be a stain on Soddy's reputation. Unexpectedly, Soddy obtained a paradoxical experimental result: Antonoff and Fleck were both right and wrong at the same time!

As a matter of fact, Antonoff was correct in confirming the existence of the new beta-emitting radioelement which he called uranium-Y. However, he was undoubtedly mistaken in his assumption that uranium-Y might be separated from uranium-X by any chemical method.

Fleck was correct in believing that chemical separation of uranium-Y from uranium-X was impossible. However, Fleck's statement about the nonexistence of uranium-Y was incorrect.

Soddy demonstrated that the existence of uranium-Y might be detected by a study of the disintegration curves of uranium-X *immediately after* its separation from uranium. Both uranium-Y and uranium-X are isotopes of thorium (thorium-231 and thorium-234, respectively) and are not separated chemically from each other. Soddy's article with this result appeared in the January, 1914 issue of the *Philosophical Magazine* [41].

Of course, after this Rutherford experienced a feeling of relief. In his letter to Georg Hevesy of January 7, 1914 he wrote:

You will have seen in this month's *Phil. Mag.* that Soddy has confirmed Antonoff's results. I am very glad that things have turned out all right, for I should have been very sorry to feel that the laboratory had one piece of bad work to its discredit. I think Antonoff has scored heavily in the matter, for he never wavered in his certainty that he was right . . . [42].

Antonoff wrote to Rutherford on January 20, 1914:

Dear Professor Rutherford,

I have just seen the January number of the *Phil. Mag.* and found out that Soddy has confirmed my main results and differs from me only in some details. In a paper sent to *Le Radium* I have given some explanations concerning these points which will show that there is almost complete agreement between my results and his.

This paper of mine will be published in the December number of *Le Radium*, which unfortunately will appear later than the January number of the *Phil. Mag.*

For that reason it would be better perhaps, to write another short Letter to the Editors of the *Phil. Mag.* Otherwise, English readers will think that I recognize some mistakes in my work. If you think that it is necessary to do so, please let me know.

On the whole there is nothing in Soddy's paper to which I can object, and the personal tone of his paper seems to be very decent, whereas Fleck was rude to me.

I have written a letter to him, as you suggested, and he did not even answer [43].

Two weeks later, on February 6, 1914, at the Annual Meeting of the Chemistry Section of the Russian Physical-Chemical Society, Antonoff informed Russian scientists about the scientific controversy concerning the discovery of uranium-Y. He first noted his previously successful repetition of experiments on separating uranium-Y from Soddy's purified uranium nitrate, and he then spoke of Soddy's confirmation of the existence of uranium-Y [44].

It appears that uranium-Y was the first radioelement, isotopic with thorium and uranium-X, to be experimentally discovered immediately after Soddy's formulation of the concept of isotopy in *Nature* on December 4, 1913 [45], because Soddy's article 'The Existence of Uranium-Y' was dated December 10, 1913 [41].

The echo of this controversy, as the present author showed earlier [46], may resound in the study of the history of the discovery of protactinium. Soddy highly estimated Antonoff's work on separating uranium-

Y as well as his original idea that this radioelement might belong to the actinium series. It is worth mentioning that Soddy and Cranston's original article on the parent of actinium concludes as follows:

The work was undertaken to test and confirm the view that the parent of actinium occupies the eka-tantalum place in the Periodic Table and gives actinium in an $\alpha$-ray change of long period, itself being formed as the product of uranium-Y, discovered by Antonoff, who suggested that it was the first member of the actinium series. But the mode of origin of actinium, though at present the most probable, has not yet been conclusively established to the exclusion of all the other possible modes of origin, discussed in the historical introduction [47].

It is interesting that the last words in this history belong to Antonoff. In 1945 he published a short article entitled 'On the Position of Uranium-Y in Radioactive Series', in which he showed that uranium-Y itself is the first disintegration product of uranium-235 [48].

In what follows we present the statements of V. I. Vernadsky, L. A. Chugaev, and M. Gor'kiĭ about Soddy and his work.

## 3. Vladimir Ivanovich Vernadsky and Soddy

Vernadsky's contributions to the development of Russian studies on radioactive phenomena have been frequently mentioned in the scientific literature. Therefore we shall mention only Vernadsky's evaluation of Soddy's works [8, 49].

We have found in Vernadsky's scientific works references (always to the original articles) both to Soddy's papers co-authored with E. Rutherford, W. Ramsay, R. Pirret, A. F. R. Hitchins, and to his own publications – five papers.

In his book *Essays in Geochemistry* Vernadsky gives fourteeen references to the corresponding places in Soddy's books [50] (Vol. 1, p. 389). It would be interesting to note that in Vernadsky's evaluation of Soddy's works one can sense an approach of a practicing contemporary rather than the accuracy and the reserve of an historian – all this notwithstanding Vernadsky's wish to estimate the historical significance of Soddy's works that he considered. Obviously, for Vernadsky Soddy was one of the main leaders of the evolving theory of radioactivity – the stamp of this perception marks all of Vernadsky's evaluations and statements concerning Soddy.

Thus, when speaking about the development of the theory of radioactivity, Vernadsky said: 'It is now possible to elucidate the movement of thought that resulted in the new understanding of the world, since three out of its four creators – E. Rutherford, F. Soddy, and M. Curie-Sklodovskaya have frequently presented, from the distance of the years, the history of formulation of the new conception [Here Vernadsky refers to Soddy's book [51]] [50] (Vol. 1, p. 344). In a number of cases Vernadsky compared several descriptions of the same discovery by various authors, as well as those of its consequences, which allowed the scientist to come up with his own point of view. For example, when presenting the history of the discovery of cosmic rays, Vernadsky says: 'A correct description of the effect of this discovery is given by Soddy' (in reference [51]) [50] (Vol. 1, p. 345).

An interesting example of such comparison of two different editions of Soddy's own books as well as the conclusion about the extremely rapid development of the theory of radioactivity can be found in that section of *Essays in Geochemistry* where Vernadsky discusses the possibility of universal radioactivity and the insufficient sensitivity of the methods used:

> A phenomenon can take place on an enormous scale, but we do not notice it when we apply our modern techniques. This has been known for a long time. It was already pointed out by F. Soddy in 1909 in his remarkable book on radioactivity [52], which exercised an enormous influence on contemporaries. He specified a limit below which the radioactivity, even if it were strong ($\alpha$-particles with the free path of 2000 cm $s^{-1}$), was not detected by the methods of that time. As one can see from a comparison of this book with its present edition (1932), the daring has advanced very far and the understanding of the world has changed in a direction quite unanticipated by the very creators of the events that were being experienced [50] (Vol. 1, pp. 235–6) [51].

Soddy's contribution to the development of the law of radioactive displacement and his introduction of the conception of isotopy was also spoken highly of by Vernadsky: 'Soddy has formulated the concept of isotopes and penetrated the phenomenon most deeply. A radiochemistry was created that was generalized for the first time by Soddy' [50] (Vol. 1, pp. 228–9).

It would be of interest to discuss examples of the use of Soddy's results by Russian scientists as the basis for their own studies. Vernadsky approved an attempt of this kind by a well-known radiochemist: 'I. F. Starik made use of Soddy's argument to advance a hypothesis which deserved serious attention and which was quite reasonable according to calculations, to the effect that lead of atomic weight 204 represents a final product of thorium decay, a decay that lasts for $10^8$ years' [50] (Vol. 1, p. 690).

Apparently, no less interesting are those cases in which Vernadsky cites in his works some of Soddy's attention-catching, brilliant examples: 'Even Soddy noticed that the atoms of a fast-decaying or of still more radioactive thoron or actinon cannot accumulate in any considerable amount. Five hundred cubic centimeters filled by them would have destroyed the whole environment and resulted in a catastrophe on the Earth' [50] (Vol. 1, p. 691).

Vernadsky also mentioned Soddy's works in his important paper 'On the Study of Radioactive Minerals of the Russian Empire' [50] (Vol. 1), which survived three editions (1910, 1911, and 1914), as well as in his book *The Experience of Descriptive Mineralogy* [50] (Vol. 2). In his paper 'On New Tasks in the Chemistry of Life', Vernadsky, having spoken of H. Moseley's scientific merits and his early demise, evaluated Soddy's scientific significance as highly as Moseley's: 'The history of F. Soddy's scientific work gives us a brilliant example of the same significance of an individual human personality, its enormous value as was the history of Moseley, who died tragically. We are indebted to him for the introduction of two completely new conceptions into human thought' [50] (Vol. 5, p. 218).

## 4. Lev Aleksandrovich Chugaev [52a] and Soddy

The first issue of the series 'Text-books for the Study of Chemistry' published by the publishing house 'Education' consists of a monograph by the St. Petersburg University Professor L. A. Chugaev, *The Periodic System of the Chemical Elements* [53]. (In their second issue of the series, 'Education' published Soddy's book *The Chemistry of the Radio-elements* [20], as we have already seen.) In the tenth chapter of his monograph, 'Radioactive Elements', Chugaev used unchanged the table representation of three radioactive series from the yearly review of Soddy's works on radioactivity for 1911 [53] (pp. 232–3) [54].

Such a small amount of attention given by Chugaev in this monograph to Soddy's works is caused first, by the brevity of the tenth chapter, and second, by the nature of Chugaev's presentation in this chapter where he frequently discussed important theoretical onceptions without mentioning the names of their authors.

Soddy's works were considered much more completely in Chugaev's last monograph *The Nature and Origin of the Chemical Elements from the Point of View of the Latest Studies of Atomic Decay and Isotopy* [55], which was published posthumously. According to contemporary history of science, Chugaev was not always accurate when he gave the dates of publication of the most important generalizations (e.g., Chugaev wrote that the 'remarkable' law of radioactive displacement was first formulated by Soddy in 1911 'and after that by Fajans' [55] (p. 62), but at the same time he always correctly presented the essence of the question under consideration. In the chapter 'The Soddy-Fajans Law' [55] (pp. 62–3), Chugaev, after having explained the meaning of the law of radioactive transformations and after having illustrated it with an example of the uranium family, remarks: 'Soddy's law was confirmed in all cases of radioactive transformations' [55] (p. 63). In another chapter, 'The Theory of Isotopy' [55] (pp. 69–73), Chugaev, referring to Soddy's monograph *The Chemistry of the Radioelements* [20], wrote that 'as early as 1911, the well-known British radiochemist F. Soddy advanced a new original conception about isotopy, which radically changed the entire appearance of the periodic table' [55] (p. 69). He gives the names of A. M. Butlerov and W. Crookes as the predecessors who closely approached the idea of isotopy on the basis of speculative constructions; according to Chugaev, Crookes' ideas 'may truly be called prophetic. Soddy, by comparison, made only one step forward when he assumed the existence of isotopes which are generally not separable from each other by means of conventional methods' [55] (p. 71).

One comes across Soddy's name in Chugaev's monograph several more times in the discussion of the results of studying isotopy among the radioactive elements. For instance, Chugaev remarked: 'Initially, the isotopes became known among the radioelements, and this led Soddy to the creation of the very theory of isotopy' [55] (p. 74). Chugaev devoted special sections of his book to the analysis of various aspects of isotopy: 'Isotopy among the Radioelements' (pp. 74–78), 'The Theory of Isotopy as a New Tool of Basic Research in the Fields of Chemistry and Radiochemistry' (pp. 78–80), 'Radiothorium and Mesothorium' (pp. 81–82), 'The First Cases of Isotopy among Non-Radioactive Radioelements. The Lead Group' (pp. 82–86), 'Isotopy as the General Property of Chemical Elements. Aston's Works. Conclusion' (pp. 90–103). Interestingly enough, in his book Chugaev gives examples of the synonymous use of concepts which are, strictly speaking, different (Section

IV): the law of radioactive displacement (in Chugaev's terminology, the 'Soddy-Fajans Law') [55] (p. 62) and the rule of radioactive displacement (in Chugaev's terminology, 'the rule of Soddy-Fajans') [55] (p. 79). Apparently, this circumstance was caused by the widespread practice of confusing these notions during the end of the second and the beginning of the third decades of the twentieth century. Also note Chugaev's reference to Soddy's forecast of the existence of uranium $X_2$, as well as a rather free exposition by Chugaev of a fragment from Aston's book [56]:

When J. J. Thomson reported the complexity of neon at the Congress of the British Association in Birmingham in 1913, Soddy immediately stressed the fact that neon and metaneon are undoubtedly isotopic, and he pointed out the enormous importance of this observation, since it provides evidence of the spread of isotopy far beyond the limits of the radioelements. The validity of his words was immediately confirmed, brilliantly and completely [55] (p. 89).

### 5. Maxim Gor'kiĭ (A. M. Peshkov) and Soddy

The author was assisted in his work on this section by V. V. Gnevashev (Department of A. M. Gor'kiĭ's Creativity, The Institute of Scientific Information on Social Sciences of the U.S.S.R. Academy of Sciences) and I. V. Distler (A. M. Gor'kiĭ Archives, The Institute of World Literature of the U.S.S.R. Academy of Sciences).

According to the materials presently available, Gor'kiĭ was acquainted with three of Soddy's popular science books: *Radioactivity*, *The Clue to Radium*, and *Science and Life* (a copy of which was sent to Gor'kiĭ by H. G. Wells), but if we take into account Gor'kiĭ's great interest in these books, we can assume that he familiarized himself with other Russian translations of Soddy's books, e.g., *Matter and Energy* [21, 22] and *The Chemistry of the Radioelements* [20].

However, the first time that Gor'kiĭ came across the phenomena of radioactivity and the most important scientists in this field was apparently before his reading of Soddy's books, namely, in 1906, when Gor'kiĭ visited Marie Curie's laboratory in the spring of 1906 and Rutherford's laboratory in Montreal in the autumn of the same year [57].

In his memoirs on Gor'kiĭ, K. Fedin wrote:

He was telling me with genuine gratitude about the Curies whom he had met personally and whose laboratory he had visited when he lived in France. When he understood from our conversation that I was somewhat familiar with chemistry, he immediately dropped the popular language and began to talk beautifully in detail about the radioactivity of the elements [58] (p. 84).

It is also known that Gor'kiĭ later discussed radioactive phenomena with the students of the Capri school [57] (p. 62).

Apparently, in 1908 Gor'kiĭ read Soddy's book *Radioactivity: An Elementary Treatise from the Standpoint of the Disintegration Theory* [13]. On May 15, 1908 he wrote to his friend K. P. Pyatnichkii:

I am overcome by ideas that are perhaps naive and funny, but more and more they are attractive to me. For example, I believe that an idea is a type of matter or, more accurately, a type of emanation of matter. That an idea and a will are identical.

I draw some conclusions – very astonishing ones – and I remember that they may be funny for someone else.

But I am not merely a daydreamer: thus Dr. Kotik, in his book on 'An Emanation of Psychophysical Energy,' gives me some support. He is not the only one; a few days ago I had read an amazing book by Soddy on radioactivity [59].

Gor'kiĭ's interest was also aroused considerably by Soddy's book *Radium and its Clue*, which he read in D. D. Khmyrov's translation [15] and which he regarded highly. For instance, in October, 1911 Gor'kiĭ wrote to A. V. Amfiteatrov: 'Do you want to experience a great pleasure? Read Soddy's book *Radium* published by "Mathesis"' [60]. In the same period Gor'kiĭ wrote to N. E. Burenin: '. . . And read Soddy's book *Radium* published by "Mathesis": it's a wonderful thing, and this is necessary to know' [59] (Vol. 29, p. 205). After a few months Gor'kiĭ wrote to O. P. Runova: 'And still the world is marching forward! I would very much like you to look at the last page of Soddy's book *Radium*; please, become acquainted with it; it is so inspiring!' [61].

When recommending Soddy's book *Radium*, among other books, Gor'kiĭ wrote to I. M. Kasatkin on October 25, 1912:

These books will introduce you to the most interesting and revolutionary sphere of contemporary experimental science – the field of the new theory of matter. The knowledge of just this discipline is fundamental and is quite necessary for a thinking person of our time. A new science is being born, which is neither physics nor chemistry but chemo-physics, and it will turn all the hypotheses of psychology upside down [62] (pp. 139–40).

Gor'kiĭ advised his young correspondent to escape the influence of individual authorities:

Read books; each of them will tell you how and about what the world thinks: not Le Bon, not Soddy, not man in general, but just the world. Since each book represents the quintessence of the world work of the human spirit, the book is the last word of many minds which has been said by one mind [62] (p. 140).

In 1920, at Gor'kiĭ's request, H. G. Wells sent thirteen scientific and popular science books (the first portion) for the House of Scientists. As the first book in his list of those that he sent to Gor'kiĭ, Wells named Soddy's work *Science and Life* [62] (p. 205). In his reply of December 3, 1920 Gor'kiĭ wrote:

Dear Wells: Up to this time I have received from you thirteen volumes of various scientific books, and the House of Scientists as well as I are very grateful to you for this present. All the books have already been reviewed by various professors; their reviews have been presented publicly during one of the Saturday meetings of scientists; some books, and first of all, the *papers and speeches of Soddy* [italics added], are being translated into Russian [59] (Vol. 29, p. 395).

Apparently, in 1922 Gor'kiĭ attempted to contact Soddy in order to obtain a paper from him. Gor'kiĭ wrote to England about this to his translator S. S. Kotelyansky, who replied on June 3, 1920: 'It seems to me that if you write to Prof. Soddy and ask him to compose a paper in his field (radioactivity) or to recommend a competent scientist, he would do this for you since he expressed very warm feelings towards you when I met him last year at Wells' house' [63]. In his letter to the members of the Gor'kiĭ Foundation, dated May 18, 1922, Gor'kiĭ wrote: 'Science, which is trying to improve the life of people, formulates ever more daring and fantastic tasks for itself, and it solves these problems boldly.... The great investigators of radioactivity, Rutherford, a former professor at Montreal, and the wonderful Soddy work continuously' [64].

The last time that Gor'kiĭ mentioned Soddy, as far as we know, was at the beginning of 1923, when Gor'kiĭ was busy with the organization of the new scientific-humanist journal *Beseda*, the goals of which he described in his letter to S. N. Sergeev-Tsensky: 'Yes, *Beseda* has nothing to do with politics, publishes no political papers. Only the positive sciences, the history of literature, poetry, and fiction. Twenty-five lists in two months, and in the autumn the journal will be made into a monthly' [59] (Vol. 29, p. 411). In order to obtain the necessary papers, Gor'kiĭ wrote to H. G. Wells on February 21, 1923:

Are you familiar with Soddy and Rutherford? Would they write small essays on the modern state of research on radium or on the atomic theory? If you do not know them, please give me their addresses. I apologize for bothering you, but I want very much to give my country an opportunity of becoming acquainted with the intellectual life of Europe, with which Russia has no ties.... Thus, may I hope to get a paper from you? Please excuse my being so obtrusive. It is not of myself that I am thinking [65] (p. 73).

Obvously, Gor'kiĭ was so sure that Wells would support his efforts that the next day, February 22, 1923, he wrote to V. M. Alexeev: '... the second book is going to contain the papers of Wells and Soddy' [59] (Vol. 29, p. 409). Apparently, because of this sentence of Gor'kiĭ's a commentator of Gor'kiĭ's and Wells' correspondence made the following remark: 'In 1923 A. M. Gor'kiĭ invited Soddy to collaborate in the journal *Beseda*' [65] (p. 78). In the preparation of the present section, the author has examined the issues of *Beseda* kept in the Department of Gor'kiĭ's Creativity, but no papers of Soddy or Wells were found. It is known that Wells sent his book *People as Gods* to Gor'kiĭ for translation, but the latter refused to publish a translation of this book in *Beseda* because this translation had appeared earlier in the weely *Krasnaya niva*, as Gor'kiĭ mentioned in his reply to Wells: 'It would not make sense to publish one and the same book simultaneouly in two different translations for the same readers' [55] (p. 74).

Undoubtedly, the fact of Soddy's being elected as foreign member-correspondent to the USSR Academy of Sciences constituted the highest evaluation of his works in Russia. Obviously, the numerous translations of his books into Russian discussed above played a considerable role in his being unanimously elected.

## 6. Soddy's Works on Economics

While preparing materials for a scientific biography of Soddy [66], the present author found that Soddy's economic ideas did not elicit any appreciable response in the USSR although there are such interesting books by Soddy on socio-economic problems as *Money versus Man* [67] and *The Inversion of Science and a Scheme of Scientific Reformation* [68] in the V. I. Lenin Moscow Central Library. As is generally known, Soddy turned to socio-economic problems during World War I (Soddy himself noted that Henry Moseley's tragic death gave rise to his mode of thinking.).

From this point of view, it is very interesting that in 1934 in Riga (now the capital of the Latvian SSR) there was published a book that was most original for that time. This book [69] was devoted to the popularization of Soddy's socio-economic views. Unfortunately, we have not been able to locate any biographical information on the author of this interesting book – Nikolaus Freiherr von der Brüggen. Therefore we can give only a short description and evaluation of its contents.

From our point of view, an obvious social need caused by the deep economic slump that gripped the Western European countries and America in the early 1930s contributed to the appearance of this book. At that time new ideas in economics were of great interest. It is therefore not surprising that Soddy's books and booklets on economics were reedited many times in the United States, although in the United Kingdom, a country of more established traditions, Soddy's new ideas were received rather coldly. It was the success of Soddy's publications in the United States that seemed to inspire von der Brüggen's introduction to his economic ideas.

Despite the fact that the book was published in 1935, it had been completed by von der Brüggen by the beginning of 1934. Soddy read the manuscript in February, 1934 and wrote quite a favorable preface, dated February 23, 1934. In this preface Soddy characterized von der Brüggen as a talented author who had contributed much work and intellectual effort to the analysis of his (Soddy's) economic works, managing to give a brief and original review of his main outlooks on the achievement of 'prosperity-for-all.' In his short preface, dated March, 1934, von der Brüggen noted Soddy's 'physical approach' to economic problems. The book contains 102 pages; it is divided into 44 short parts, which discuss various aspects of Soddy's economic ideas (money and credit systems; technocracy; physics versus psychology; prosperity; the laws of thermodynamics; fictitious money; personality versus society; banks and bankers; the check system; inflation, etc.).

As a whole, the book appears very interesting, but we have no information on the success of this edition, the number of its copies, or its future editions.

## 7. Soddy and the U.S.S.R. Academy of Sciences

In 1924 the Academy of Sciences of the U.S.S.R. elected among the number of its honorable member-correspondents many outstanding foreign scientists, including Francis W. Aston, Niels Bohr, Max Born,

M. Wolf, Peter Debye, W. Campbell, Paul Langevin, Max von Laue, Albert A. Michelson, Robert Millikan, Jean Perrin, Ernest Rutherford, Frederick Soddy, J. J. Thomson, Kasimir Fajans, and Paul Ehrenfest [70].

During the twelfth session of the Russian Academy of Sciences (December 6, 1924), the foreign member-correspondents, including Soddy, elected to the various sections (sciences) were approved. The 1924 volume of *Izvestiya of the Russian Academy of Sciences* contains a brief 'Note about the Scientific Works of F. Soddy', signed by the academicians N. Kurnakov, D. Konovalov, V. Ipat'ev, P. Lazarev, and A. Ioffe [71].

In the Leningrad section of the U.S.S.R. Academy of Sciences Archives we discovered two of Soddy's letters:

April 1, 1925

Dear Sir,

Your communication of February 28th which I have just received, informing me that l'Académie des Sciences de Russie have done me the honour of adding my name to the list of their corresponding members, has given me very great pleasure. Will you kindly convey to the members of the Academy my sincere appreciation and grateful thanks for thus associating me with their illustrious body – Believe me.

Yours very sincerely,
Frederick Soddy [72]

March 21, 1926

Dear Sir,

Thank you for the Diploma you have sent me signalling my election as corresponding member of the Academy of Sciences of Russia.

Yours sincerely,
Frederick Soddy [73]

In the list of documents kept in the Soddy Archives at the Bodleian Library at Oxford University we have not found the corresponding notifications of the Secretary of the U.S.S.R. Academy of Sciences about Soddy's election as a member-correspondent of the Academy of Sciences of the U.S.S.R. (Apparently, these documents are being kept in other archives.); this list, however, contains an invitation issued to Soddy to visit Moscow and the U.S.S.R. Academy of Sciences in 1945 [74] (p. 26, No. 249).

## 8. The Analysis of Soddy's Scientific Heritage by Soviet Historians of Science

Of Soddy's vast scientific heritage, Soviet historians of science have studied mainly his prewar works on radioactivity (those published before World War I), with the greatest degree of attention being paid to his works with Rutherford and Ramsay as well as to his contribution to the discovery of the law of radioactive displacement and of the phenomenon of isotopy. Soddy's economic and social views have not been studied at all in our country.

B. M. Kedrov [3, 75] was first among the Soviet historians of science who began to analyze Soddy's contribution to the discovery of the law of radioactive displacement and of the phenomenon of isotopy. Many of Kedrov's generalizations are still not out of date today, for instance, the following conclusion: 'The displacement law emerged as a new concrete formulation of the periodic law applied to the phenomena of radioactivity' [3] (p. 159). A valuable study of the Rutherford-Soddy collaboration has been carried out by O. A. Starosel'skaya-Nikitina [76, 77]. L. L. Zaitseva and N. A. Figurovsky made a well-documented study of the argument between G. N. Antonoff and Soddy's school, which was concerned with the reliability of the discovery of the new radioelement uranium-Y [8] (pp. 111–4). The works that have been written recently by scientists from the Group of the History of the Theory of Periodicity in the Institute for the History of Science and Technology of the USSR Academy of Sciences, Moscow consider on a much broader scale than before Soddy's contribution to the theory of radioactivity and the theory of periodicity [78–88]. The authors of this group plan to study further Soddy's scientific heritage. The first publications in Russian of Soddy's scientific biography and the translation and publication of his last popular science book, *The Story of Atomic Energy*, were published in 1978 [66] and 1979 [82].

It is my pleasure to thank the scientists of the Sector of the History of Chemistry in the Institute for the History of Science and Technology, Moscow for their valuable discussions of the present paper. My special gratitude is expressed to Dr. D. N. Trifonov and Prof. Yu. I. Solov'ev.

*Institute for the History of Science and Technology*
*Academy of Sciences of the USSR*
*Staropanskiĭ per. 1/5*
*Moscow K-12, USSR*

## Bibliography

[1] Brock, W. H.: 1969, 'Lockyer and the Chemists: The First Dissociation Hypothesis', *Ambix* **16**: 1 and 2, 81–99.

[2] Meadows, J. A.: 1972, *Science and Controversy: A Biography of Sir Norman Lockyer*. London: Macmillan.

[3] Kedrov, B. M.: 1956, *Evol'yuciya ponyatiya elementa v khimii*. Moscow: izdatelstvo APN USSR.

[4] Brock, W. H.: 1965, 'The Life and Work of William Prout', *Medical History* **9:** 2, 101–26.

[5] Solov'ev, Yu. I.: 1971, *Evol'yuciya osnovnykh teoreticheskikh problem khimii*. Moscow: Nauka.

[6] Makareniya, A. A.: 1975, *D. I. Mendeleeff o radioaktivnosti i slozhnosti elementov*, 3th ed. Moscow: Atomizdat.

[7] Rodnyi, N. I. and Solov'ev, Yu. I.: 1969, *Wilgelm Ostwald (1853–1932)*. Moscow: Nauka.

[8] Zaitseva, L. L. and Figurovskii, N. A.: 1961, *Issledovaniya yavlenii radioaktivnosti v dorevolyutsionnoi Rossii*. Moscow: izdatelstvo AN USSR.

[9] Ramsay, W. and Soddy, F.: 1903, 'Poluchenie geliya iz radiya', *Fizicheskoe obozrenie* 253–256.

[10] Giesel, F.: 1903, *O radioaktivnykh veshchestvakh i ikh luchakh*. Per. A. E. Chichibabina. Pod red. prof. M. I. Konovalov. Moscow: tipography tovarishchestva I. D. Sytina.

[11] Hofmann, K. A.: 1903, *Radii i yego luchi*, Per. F. N. Indriksona. Pod red. prof. I. I. Borgmana. St. Petersburg: tipografiya M. M. Stasyulevicha.

[12] Soddy, F.: 1903, *Radioaktivnye yavleniiya* (*Iz zhurnala 'Elektrichestvo'*). St. Petersburg: tipografiya M. Merkusheva.

[13] Soddy, F.: 1905, *Radioaktivnost. Elementarnoe izlozhenie s tochki zreniya teorii raspada atomov*, Per. F. N. Indrikson. St. Petersburg: tipografiya M. M. Stasyulevicha.

[14] Soddy, F.: 1910, *Radii. Obshchedostupnye lektsyi o prirode radiya, chitannye v universitete v Glasgo v 1908*, Per. N. A. Shilov. Moscow: tipo-litografiya tovarishchestva I. N. Kushnerev i Ko.

[15] Soddy, F.: 1910, *Radii i yego razgadka*, Per. pod red. D. D. Khmyrova. Odessa: Mathesis.

[16] Soddy, F.: 1915, *Radii i yego razgadka*, Per. pod red. D. D. Khmyrova. Odessa: Mathesis.

[17] Soddy, F.: 1923, *Radii i stroenie atoma*, Per. pod red. D. D. Khmyrova. 3th ed. Odessa: Mathesis.

[18] Soddy, F.: 1924, *Radii i stroenie atoma*, Per. i predislovie Prof. N. A. Shilova. Moscow: Gosudarstvennoe izdatelstvo.

[19] Soddy, F.: 1924, *Radii i yego razgadka*, Per. E. A. Tolmachovoi. Pod red. V. G. Khlopina. Leningrad: Nauchnoe khimiko-tekhnicheskoe izdatelstvo.

[20] Soddy, F.: 1913, *Khimiya radioelementov*. Per. R. Shmidt. Pod red. i s primechaniyami V. A. Borodovskogo. St. Petersburg: Obrazovaniye.

[21] Soddy, F.: 1913, *Materiya i energiya*, Per. S. G. Zaimovskogo. Pod red., s predisloviem i primechaniyami Nikolaya Morozova. Moscow: Priroda.

[22] Soddy, F.: *materiya i energiya*, Per. s angl. K. A. Leont'eva. Pod red. A. K. Timiryazeva. Moscow: Pechatnik.

[23] Soddy, F.: 1909, 'The Production and Rays of Uranium X', *Nature* **79**, 366–367.

[24] Soddy, F.: 1909, 'The Parent of Radium', *Scientia* **5**, 256–276.

[25] Soddy, F.: 1909, 'The Rays of Uranium X', *Nature* **80**, 37–38.

[26] Soddy, F.: 1909, 'Production of Radium from Uranium'. *Nature* **80**, 308–309.

[27] Soddy, F.: 1909, 'The Relation between Uranium and Radium, Part IV', *Phil. Mag.* **18**, 846–858.

[28] Soddy, F.: 1909, 'The Rays and Product of Uranium X', *Phil. Mag.* **18**, 858–865.

[29] Soddy, F.: 1910, *Radioactivity. Annual Progress Report to the Chemical Society for 1908/09*. London: Taylor & Francis Ltd., Vol. 6, pp. 232–267.

[30] Antonoff, G. N.: 1911, 'The Disintegration Products of Uranium', *Phil. Mag.* **22**, 419–432.

[31] Badash, L.: 1969, *Rutherford and Boltwood: Letters on Radioactivity*. New Haven: Yale University Press, p. 250.

[32] Fleck, A.: 1913, 'The Existence of Uranium Y', *Phil. Mag.* **26**, 710–712.

[33] Fleck, A.: 1913, 'The Disintegration of Uranium X', *Phil. Mag.* **26**, 529–535.

[34] Antonoff, G. N.: 1913, *Produkty dezintegratsii urana. Issledovaniya iz oblasti radioaktivnykh yavlenii*. St. Petersburg: Ekonomicheskaya tipo-litografiya, p. 99.

[35] Antonoff, G. N.: 1913, 'The Existence of Uranium Y', *Phil. Mag.* **26**, 332–333.

[36] Rutherford MS Collection, Cambridge Univ. Library, Add 7653/S 167.

[37] Rutherford MS Collection, Cambridge Univ. Library, Add 7653/A 51.

[38] Antonoff, G. N.: 1913, 'Discussion', *Zhurnal russkogo fiziko-khimicheskogo obshchestva* **45**: 2, 1463–1464.

[39] Rutherford MS Collection, Cambridge Univ. Library, Add 7653/S 169.

[40] Antonoff, G. N.: 1913, 'On the Existence of Uranium Y', *Phil. Mag.* **26**, 1058.

[41] Soddy, F.: 1914, 'The Existence of Uranium Y', *Phil. Mag.* **27**, 215–221.

[42] Eve, A. S.: 1939, *Rutherford. Being the Life and Letters of the Rt. Hon. Lord Rutherford, O. M.* Cambridge: At the University Press, p. 228.

[43] Rutherford MS Collection, Cambridge Univ. Library, Add 7653/A 52.

[44] Antonoff, G. N.: 1914, 'Discussion', *Zhurnal russkogo fiziko-khimicheskogo obshchestva* **46**: 1, 171.

[45] Soddy, F.: 1913, 'Intra-atomic Charge', *Nature* **92**, 399–400.

[46] Krivomazov, A. N.: 1976, 'Problemy prognozirovaniya na rannikh etapakh razvitiya ucheniya o radioaktivnosti', in *Problemy prognozirovaniya v khimii*. Moscow: Nauka Press, 224–247.

[47] Soddy, F. and Cranston, J. A.: 1918, 'The Parent of Actinium', *Proc. Roy. Soc.* **94A**, 384–404; p. 404.

[48] Antonoff, G. N.: 1945, 'On the Position of Uranium Y in Radioactive Series', *Phys. Rev.* **68**, 288.

[49] Pogodin, C. A. and Libman, E. P.: 1971, *Kak dobyli sovetskii radii*. Moscow: Atomizdat.

[50] Vernadsky, V. I.: 1958–1960, *Izbrannye sochineniya v 5 tomakh*. Moscow: izdatelstvo AN USSR.

[51] Soddy, F.: 1932, *The Interpretation of the Atom. Part I. The Radioactive Elements and Isotopes; Part II. The General Progress of Atomic Chemistry*. London: John Murray.

[52] Soddy, F.: 1909, *The Interpretation of Radium*. London: John Murray.
[52a] Kauffman, G. B.: 1963, 'Terpenes to Platinum: The Chemical Career of Lev Aleksandrovich Chugaev., *J. Chem. Educ.* **40**, 656–665.
[53] Chugaev, L. A.: 1911, *Periodicheskaya sistema khimicheskikh elementov*. St. Petersburg: Obrazovanie.
[54] Trenn, T. J. (ed.): 1975, *Radioactivity and Atomic Theory*, London: Taylor & Francis Ltd., pp. 253–688.
[55] Chugaev, L. A.: 1923, *Priroda i proiskhozhdenie khimicheskikh elementov v svyazi s noveyshimi issledovaniyami o raspade atoma i ob isotopii*. Petrograd: Nauchnoe khimiko-tekhnicheskoe izdatelstvo.
[56] Aston, F. W.: 1923, *Isotopy*. Odessa: Mathesis.
[57] Yunovich, M. M.: 1961, *M. Gor'kiĭ – propagandist nauki*. 2nd ed. Moscow: Sovetskii pisatel.
[58] Fedin, K.: 1943, *Gor'kiĭ sredi nas. Dvadtsatye gody*. Moscow: Goslitizdat.
[59] Gor'kiĭ, M.: 1955, *Sobranie sochinenii v 30 tomakh*. Tom 29. Moscow: Khudozhestvennaya literatura, p. 71.
[60] A. M. Gor'kiĭ Archives, PG-rl 1-25-129.
[61] A. M. Gor'kiĭ Archives, PG-rl 37-23-2.
[62] *Gor'kiĭ i nauka, Stat'i, rechi, pisma, vospominania*. Moscow: Nauka, 1964.
[63] A. M. Gor'kiĭ Archives, KG-P 39-9-1.
[64] A. M. Gor'kiĭ Archives, PG-Kou 64-24.
[65] *Perepiska A. M. Gorkogo s zarubezhnymi literatorami*. Arkhiv A. M. Gorkogo. Vol. VIII. Moscow: izdatelstvo AN U.S.S.R., 1960.
[66] Krivomazov, A. N.: 1978, *Frederick Soddy (1877–1956)*. Moscow: Nauka.
[67] Soddy, F.: 1931, *Money Versus Man. A Statement of the World Problem from the Standpoint of the New Economics*. London: Mathews and Marrot.
[68] Soddy, F.: 1924, *The Inversion of Science and a Scheme of Scientific Reformation*, London: Henderson.
[69] Brüggen, Nikolaus Freiherr von der: 1934, *Wohlstand für Alle. Grundsätzliche Orientirung über eine Neue Nationalökonomie nach Professor Frederick Soddy, M. A. (Oxon.), F. R. S. Nobelpreisträger in Chemie*. Riga: E. Bruhns.
[70] Leningrad Section, Archives of the USSR Academy of Sciences, Fund No. 2, Inventory No. 1 (1925), No. 16, pp. 28–29.
[71] Kurnakov, N., Konovalov, D., Ipatiev, V., Lazarev, P., and Ioffe, A.: 1924, 'Zapiska ob uchenykh trudakh F. Soddy', *Izvestiya Rossiyskoi Akademii nauk*, **181**, 12–18, 481–482.
[72] Leningrad Section, Archives of the U.S.S.R. Academy of Sciences, Fund No. 1, Inventory No. 2 (1926), No. 5/32, § 69.
[73] Leningrad Section, Archives of the U.S.S.R. Academy of Sciences, Fund No. 2, Inventory No. 1 (1926), No. 1686, p. 134.
[74] *Report on the Papers of Professor Frederick Soddy, F. R. S. (1877–1956)*. London: The Royal Commission on Historical Manuscripts, 1974.
[75] Kedrov, B. M.: 1948, *Razvitie ponyatiya elementa ot Mendeleeva do nashikh dnei*. Moscow-Leningrad: Gosudarstvennoe nauchno-tekhnicheskoe izdatelstvo.
[76] Staroselskaya-Nikitina, O. A.: 1963, *Istoriya radioaktivnosti i vozniknoveniya yadernoi fiziki*. Moscow: izdatelstvo AN U.S.S.R.
[77] Staroselskaya-Nikitina, O. A.: 1967, *Ernest Rutherford (1871–1937)*. Moscow:

Nauka.
[78] *Uchenie o radioaktivnosti. Istoriya i sovremennost.* Moscow: Nauka, 1973.
[79] *Prognozirovanie v uchenii o periodichnosti.* Moscow: Nauka, 1976.
[80] Trifonov, D. N., Krivomazov, A. N., and Lisnevskiĭ, Yu. I.: 1974, *Uchenie o periodichnosti i uchenie o radioaktivnosti. Kommentirovannaya khronologiya vazhneishikh sobytii*, Moscow: Atomizdat.
[81] Vyal'tsev, A. N., Krivomazov, A. N., and Trifonov, D. N.: 1976, *Pravilo sdviga i yavlenie isotopii.* Moscow: Atomizdat.
[82] Soddy, F.: 1979, *Istoria atomnoi energii*, Per. M. Yu. Bogdanov, M. V. Kolokol'nikov A. Yu. Pentin, Pod red., s predisloviem i s primechaniyami A. N. Krivomazov, and D. N. Trifonov. Moscow: Atomizdat.
[83] Krivomazov, A. N.: 1979, 'Wissenschaftliche Schulen: Fundamentale Entdeckungen und Fragen des Prestiges', in *Wissenschaftliche Schulen*, Band 2. Berlin: Akademie-Verlag, pp. 39–60.
[84] Trifonov, D. N., Krivomazov, A. N., and Lisnevskiĭ, Yu. I.: 1980, *Khimicheskie elementy i nuklidy. Spetsifika otkrytii*, Moscow: Atomizdat.
[85] Altshuler, S. V., Krivomazov, A. N., Melnikov, V. P., Petrov, L. P., and Trifonov, D. N.: 1980, *Otkrytie khimicheskikh elementov. Spetsifika i metody otkrytiya. Posobie dlya uchifelei.* Moscow: Prosvetchenie.
[86] Krivomazov, A. N.: 1981, 'Istoria razmetcheniya radioaktivnykh elementov v periodicheskoi sisteme', in *Uchenie o periodichnosti. Istoria i sovremennost.* Moscow: Nauka, pp. 78–116.
[87] Krivomazov, A. N.: 1981, 'The Importance of the Discovery of Radioactivity and Radium in the Context of the Scientific-Technological Progress', in *Revolutionary Changes in Science and Technology at the Turn of the 19th and 20th Centuries.* Prague: Acta historiae rerum naturalium nec non technicarum, pp. 168–186.
[88] Krivomazov, A. N.: (in press), *Radioaktivnost i radioelementy (1896–1918).* Moscow: Nauka.

MINORU TANAKA† AND KAZUO YAMASAKI

# 7. EARLY STUDIES OF RADIOACTIVITY AND THE RECEPTION OF SODDY'S IDEAS IN JAPAN

*Chemical research and education in Japan began soon after the time of the Meiji Restoration (1868) with the importation of nine British scientists and the founding of a university and a polytechnic college in Tokyo. The first radioactive mineral found in Japan (hokutōlite) was reported in 1908, and the first Japanese article on the measurement of radioactivity – of thermal spring waters – appeared in 1910, the year in which the first Japanese course on radioactivity was initiated by Suekichi Kinoshita of Tokyo Imperial University. Satoyasu Iimori, who graduated from this university in 1910, worked with Frederick Soddy at Oxford University (1920–1921) and returned to Japan in 1921 to begin a lengthy series of systematic studies that earned him the title of the founder of Japanese radiochemistry. The radiochemical research of Iimori and other Japanese scientists led directly to the thriving research now under way in post-War War II Japan.*

## 1. Early Science Education in Japan

In 1868 the Meiji Restoration restored direct Imperial rule to Japan, abolished the feudal rule of the Tokugawa family, and created a modern central state. In 1877 a university and a polytechnic college were established in Tokyo. Kobu Daigakko, the polytechnic college, was organized according to the plan proposed by the Scottish engineer William John Macquorn Rankine (1820–1872) in Edinburgh. Upon his recommendation nine British scientists were employed by the Japanese government as teachers for each department of the college. Edward Divers (1837–1912)[1] was one of them. He had studied under August Wilhelm von Hofmann (1818–1892) at the Royal College of Chemistry in London, both studied and taught chemistry in Japan from 1873 to 1899, and is considered the founder of Japanese inorganic chemistry.

Tokyo University, called Tokyo Imperial University after 1886, was established on the base of Kaiseisho, an institute for teaching and translation of western science and technology under the feudal govern-

*George B. Kauffman (ed.), Frederick Soddy (1877–1956)*, 141–154.

ment that had been founded in 1855. The institute has had a section for chemistry since 1960. On the recommendation of Alexander William Williamson (1824–1904) in London, in 1874 Tokyo University hired Robert William Atkinson (1850–1929)[2] to teach analytical chemistry and applied chemistry. Atkinson remained until 1881, and many of his students became Japan's leading applied chemists. In 1886 the poly-

Fig. 1. A bronze bust of Edward Divers. Department of Chemistry, Faculty of Science, The University of Tokyo.

technic college merged with Tokyo University to form Tokyo Imperial University, with Divers as Professor of Chemistry in the Faculty of Science. Tamemasa Haga (1856–1914),[3] who had studied under Divers at the polytechnic colelge and had been Assistant Professor in the Department of Chemistry of the Imperial University. Their lengthy collaboration on sulfur and nitrogen compounds constituted an important systematic contribution to inorganic chemistry. Haga was promoted to Professor in 1899 when Divers returned to England.

Fig. 2. Tamemasa Haga (1856–1914).

After this time education in the Department of Chemistry, Faculty of Science, Tokyo Imperial University was entrusted to Japanese professors, Jōji Sakurai (1858–1939),[4] Tamemasa Haga, Kikunae Ikeda (1864–1936),[5] and several assistant professors. Sakurai, who had studied under Williamson during his stay in London (1876–1881), was deeply influenced by Williamson's chemical thought, particularly by his atomistic thinking. Sakurai stimulated a great interest in physical chemistry and organic stereochemistry in his students. One of them, Yuji Shibata (1882–1980)[6] continued the tradition of inorganic chemistry founded by Divers and Haga by performing a series of comprehensive studies on coordination chemistry after working in the laboratories of Arthur Hantzsch (1857–1935) in Leipzig, Alfred Werner (1866–1919)[7] in Zürich, and Georges Urbain (1872–1938) in Paris. Introduction of

Fig. 3. The staff of the Department of Chemistry, Faculty of Science, Tokyo Imperial University in July, 1919. From left, Satoyasu Iimori, Yuji Shibata (Professor of Inorganic Chemistry), Kikunae Ikeda (Professor of Physical Chemistry), Jōji Sakurai (Professor of Physical Chemistry), and Kōichi Matsubara (Professor of Organic Chemistry). Other members are not shown. At this time Iimori was a member of The Institute for Physical and Chemical Research, Tokyo and gave lectures at the Department of Chemistry.

Fig. 4. Yuji Shibata (1882–1980). This photograph was taken when he gave a lecture at the Tokyo Institute of Technology on the occasion of the centennial of Alfred Werner's birth, December 12, 1966.

absorption and emission spectrographic techniques was Shibata's primary contribution to inorganic chemistry in Japan.

By using emission spectrographic analysis, Shibata initiated in 1921 his systematic studies of Japanese minerals containing the rarer elements, which constitute the beginning of geochemistry in Japan. One of Shibata's students, Kenjirō Kimura (b. 1896),[8] who had studied under

György von Hevesy (1885–1966) at Niels Bohr's (1885–1962) institute in Copenhagen (1925–1927), trained an effective group of chemists who later contributed much to the development of radiochemistry and nuclear chemistry in Japan.

Satoyasu Iimori (1885–1982),[9] who graduated from the Department of Chemistry, Tokyo Imperial University in 1910 after studying under Haga, later became the founder of radiochemistry in Japan. He was one of the first students to attend the course on radioactivity, instituted in 1910 for the Department of Physics by Suekichi Kinoshita (1877–1934),[10] who had studied in Manchester under Ernest Rutherford (1871–1937). After teaching analytical chemistry in the Department of

Fig. 5. Kenjirō Kimura (b. 1896) at the age of sixty when he retired from the University of Tokyo.

Fig. 6. Satoyasu Iimori (1885–1982).

Chemistry, Faculty of Science, Tokyo Imperial University and chemistry in the First High School, Iimori entered The Institute for Physical and Chemical Research, Tokyo when it was established in 1917 (*vide infra*). Sakurai arranged for Iimori to study under Frederick Soddy (1877–1956) at Oxford University during 1920 and 1921.

Frederick Soddy graduated from Oxford University in 1895 and worked with Rutherford on the disintegration theory of radioactive elements at McGill University, Canada during 1901–1903. At Glasgow University (1904–1914) he formulated his displacement law and the concept of isotopy. Soddy moved to Oxford University in 1919 after working at Aberdeen University (1914–1919).

Iimori arrived at Oxford in 1919 when Soddy had just moved from Aberdeen, and his laboratory was not yet arranged. Consequently, Iimori worked for one year in the laboratory of metallurgy on the gravimetric determination of arsenic. In 1920 Iimori started to work in Soddy's laboratory and determined uranium in Ceylon monazite using the emanation electrometer designed by Soddy. His study also included the application of U $X_1$ as an indicator to examine the behavior of thorium during analysis of monazite. In 1921 Soddy was awarded the Nobel prize in chemistry. In later life Soddy became interested in the economic side of human life and wrote articles on this subject.

After returning to Japan in 1921, Iimori taught a course on 'Chemistry of the Radioelements' at Tokyo Imperial University for more than twenty years (1922–1943). He chose the title for his course from Soddy's book on the subject.[11] The present authors attended his lectures when they were students of that university. Iimori worked primarily as Chief Researcher of the Rikagaku Kenkyūsho (The Institute for Physical and Chemical Research, Tokyo), which was established in 1917 on the advice of Sakurai and others and which made great contributions to the progress of physical science in Japan. Since Iimori worked in a research institute rather than a university, he did not train many students.

## 2. Early Japanese Studies of Radioactivity and Radiochemistry

In the period 1891–1899, twenty-two years after the foundation of the Chemical Society of Japan (then the Chemical Society of Tokyo), the discoveries of radium, polonium, and actinium were reported as overseas news in the Journal of the society. The first Japanese radioactive mineral was found in 1906 and reported in 1908 by its discoverer, the mineralogist Yōhachirō Okamoto (1876–1960).[12] The chemical analysis of this mineral, a barium-lead sulfate later named hokutōlite (found in the Hokutō thermal spring near Taipei, Taiwan), was performed in 1908 by Haga and his students, including Iimori.

The first radioactivity measurement of thermal spring waters in Japan was performed in 1910 by Kaichirō Manabe (1878–1941)[13] and Den-ichirō Ishitani (1877–1941),[14] using the 'Fontactoscope' of C. Engler and H. Sieveking. In 1915 Risaku Ishizu[15] of the Tokyo Institute of Hygienic Research published a book in English, 'The Mineral Springs of Japan', which included radioactivity data on almost all Japanese mineral springs known at that time.

In 1910, on the occasion of the close approach of Halley's Comet, Kinoshita collected radioactive deposits in the atmosphere on a wire 30 metres long with a high negative potential and measured the radioactivity using a self-made gold leaf electrometer of the C. T. R. Wilson type in collaboration with his students, including Iimori. No particular effect due to the comet was found.

Importation of radium preparations for medical use began in 1912 by a commercial firm in Tokyo, which preserved a standard preparation containing a radium salt corresponding to 10 mg of radium metal.

The first announcement, in 1913, of Frederick Soddy's new concept of isotopy initially aroused almost no interest in Japanese scientific circles. However, the lecture[16] that Soddy presented in December, 1918 before the Chemical Society in London on the concept of chemical elements based on his studies of radioactive disintegration was read by Iimori and 'discussed with great interest' early in 1919 in a colloquium at the Department of Chemistry, Tokyo Imperial University. In a 1959 article[17] Iimori wrote:

> It was decided to give the term *isotope*, which had attracted no interest until then, a new Japanese name, *doigenso*, which means chemical elements occupying the same position in the periodic table.... For scientific circles in Japan at that time, Soddy's lecture was nothing but a complete surprise. Thus the truly scientific interest in radioactivity in Japan was aroused after the First World War.

## 3. Systematic Radiochemical Studies in Japan

Systematic studies of radiochemistry were begun in Japan by Iimori on his return from Soddy's laboratory in 1921. In the laboratory of the Institute for Physical and Chemical Research, Tokyo Iimori developed a series of instruments for simple and accurate measurement of radioactivity, particularly for use by chemists. With these instruments and standardized radioactive preparations which he had brought from Soddy's laboratory, a systematic method was devised to make possible the analytical determination of about $10^{-12}$ g of radium and $10^{-4}$ g of thorium in 1 gram of sample. Beginning in 1922, Iimori himself spent many years in surveying uranium- and thorium-containing minerals on the mainland of Japan and Korea, and he continued this survey until the end of the Second World War in 1945.

In collaboration with one of his pupils, Kenjirō Kimura, in 1921 Shibata published his first paper[18] on 'Chemical Investigation of Japanese Minerals Containing Rarer Elements'. He was motivated for this work by Haga, who had experienced great difficulties in analyzing minerals containing the rarer elements. Shibata applied emission spectrographic analysis to such minerals. One of the first minerals studied by Shibata and Kimura was naegite from the small village of Naegi, Gifu prefecture, a variety of zircon containing rare earths, niobium, tantalum, thorium, and uranium. It was the first radioactive mineral discovered on the Japanese mainland. This series of chemical studies of minerals continued for more than 50 years since 1921; the last paper (the 66th report in the series) was published in 1973 by K. Nagashima, one of Kimura's pupils.

The chemical studies of minerals led Shibata to a new field of chemistry – geochemistry. Geochemical studies of minerals, volcanic eruptions, mineral springs, etc., all of which required radiochemical study, were developed by Shibata and Kimura in collaboration with their students. Shibata himself thought that the so-called juvenile water,[19] which may be liberated from deep magma, should contain larger or smaller amounts of heavy water than normal water found on the earth's surface. He and his collaborators tried to confirm this idea by accurately determining the densities of waters contained in various volcanic eruptions. Higher densities were generally found, but the results were too complex to be explained by such a simple idea.

Masayoshi Ishibashi (1896–1978),[20] Professor at Kyoto Imperial University, who had also studied in 1930–1932 under Hevesy in Germany, developed the field of radiochemical analysis and radioactive indicators in Japan. He published his first paper in this field in 1934.

A cyclotron was constructed at The Institute for Physical and Chemical Research, Tokyo in 1937 and another larger one in 1944. In 1940 Yosio Nisina (1890–1951)[21] of the institute, who had worked during the period 1921–1928 in Niels Bohr's laboratory in Copenhagen and who is now considered the father of nuclear physics in Japan, discovered uranum-237 by use of the cyclotron in collaboration with Kimura. In

1941 and 1942 they and their colleagues attempted unsuccessfully to discover the element with atomic number 93 (neptunium), which should have been formed through the beta-transformation of uranium-237.

## 4. Post – World War II Studies

In 1945, after the surrender of Japan, the cyclotrons of The Institute for Physical and Chemical Research, Tokyo were destroyed by the U.S. Occupation Forces. The difficult postwar period did not prevent Japanese radiochemists from carrying out useful work. From autumn to winter in 1945 comprehensive scientific surveys of damages produced by the atomic bombs in Hiroshima and Nagasaki were organized. Under extremely difficult conditions, Kimura and his collaborators carried out almost complete determinations of nuclides contained in radioactive deposits found in the two cities.[22] At that time just after the war no city gas and water were available in Tokyo Imperial University, and Kimura's laboratory had no Geiger-Müller (G.M.) counter.

In the Hiroshima deposits the presence of Sr-89, Ba-140, and La-140 was confirmed, and that of Y-91, Zr-95, and Nb-95 was suspected. In the Nagasaki deposits Sr-89, Zr-95, Ba-140, Ce-144, and Pr-144 were identified. In 1951 Kimura and his collaborators analyzed again the Nagasaki deposits which had been kept since 1945, and they identified Sr-90, Y-90, Cs-137, Ba-137, Ce-144, Pr-144, and Pu-239. The last element is presumed to be the material of the second atomic bomb, which was dropped on Nagasaki. Their radiochemical studies contributed much to medical studies of atomic bomb radiation sickness.

In 1954 a hydrogen bomb experiment was carried out by the United States in the lagoon on the Pacific atoll of Bikini. In order to save the fishermen who had received dangerous injury from the radioactive fallout deposited on their fishing boat, Daigo Fukuryū Maru (Fifth Lucky Dragon), and in order to determine the radioactive pollution of sea waters, fish, precipitated water, etc., many Japanese radiochemists concentrated their efforts. Kimura and his collaborators contributed a series of exact determinations of nuclear species contained in the fallouts, particularly in the so-called 'ash of death' that fell on the fishing boat, Daigo Fukuryū Maru.[23]

1954 was also the year of the foundation of the Institute for Nuclear Study, University of Tokyo, which was followed by the establishment of the Institute of Atomic Energy Research in 1956. With the rapid growth of atomic energy plants in Japan, many scientists engaging in atomic physics and radiochemistry are now confronting new social problems.

Iimori, who passed away in 1982 at the age of ninety-seven, had pointed out the human-welfare problems and social evils arising from the use of radioactive isotopes: 'It is our social responsibility absolutely not to entrust radiochemistry to heartless and thoughtless persons'.[24]

## Acknowledgment

The original paper of Minoru Tanaka, which had been read at the Soddy Session, International Congress on the History of Science held at Edinburgh in 1977, was revised and supplemented after his death by Kazuo Yamasaki, who is grateful to the Editor, Professor George B. Kauffman for his advice and linguistic revision of the manuscript.

## Notes

[1] Edward Divers, F. R. S. stayed in Japan from 1873 to 1899, teaching inorganic chemistry at the Polytechnic College and later at Tokyo Imperial University. *J. Chem. Educ.* **30** (1953) 234.

[2] R. W. Atkinson graduated from University College, London in 1871 and studied under Williamson. He stayed in Japan from 1874 to 1881. *J. Chem. Soc.* **1931**, 1024.

[3] Tamemasa Haga graduated in 1881 from the Polytechnic College and taught inorganic chemistry as Assistant Professor (1882–1899) and as Professor (1899–1913). He resigned one year earlier than the age limit because of illness and was succeeded by Yuji Shibata.

[4] Jōji Sakurai studied chemistry under Williamson at University College, London (1876–1881) and was Professor of Chemistry at Tokyo University, later Tokyo Imperial University (1882–1919). He advised the Japanese government to establish The Institute for Physical and Chemical Research, Tokyo, the Japan Society for the Promotion of Science (JSPS), and other societies. The grand old man of Japanese chemistry, he was made a Baron in 1939, *Nature* **144** (1939) 234; *J. Am. Chem. Soc.* **61** (1939) 2255.

[5] Kikunae Ikeda graduated from Tokyo University in 1889 and studied under Wilhelm Ostwald (1853–1932) from 1899 to 1901. In 1908 he discovered monosodium glutamate as a substance to improve the taste of foods.

[6] Yuji Shibata graduated from Tokyo University in 1907 and studied in Europe for three years (1910–1913). Considered the founder of coordination chemistry and geochemistry in Japan, he was Professor of Chemistry at Tokyo Imperial University (1919–1941) and at Nagoya University (1941–1948). He was also President of the Tokyo Metropolitan University (1949–1957) and President of the Japan Academy (1962–1970). He passed

away on his 98th birthday in 1980. For an éloge see K. Yamasaki: *J. Coord. Chem.* **11** (1981) 65. A complete list of his scientific papers was prepared and published by K. Yamasaki (*Kagaku-Shi-Kenkyū* (*J. Japanese Soc. History of Chem.*) **19** (1982) 33; **20** (1982) 97.

[7] Kauffman, George B. *Alfred Werner: Founder of Coordination Chemistry*. Berlin-Heidelberg-New York: Springer-Verlag, 1966.

[8] Kenjirō Kimura, a member of the Japan Academy, graduated from Tokyo Imperial University in 1918 and taught chemistry there (1933–1956). Later he served as the director of the Tōkai Research Institute, one of the institutes under the control of the Institute of Atomic Energy Research.

[9] Satoyasu Iimori graduated from Tokyo Imperial University in 1910. After World War II he retired from the Institute for Physical and Chemical Research, Tokyo, and he began to make artificial gem stones such as cat's eye in his own private laboratory. For éloges see Shin Hata. *Bunseki* (1983) 49; idem, *Isotope News* **343** (Jan. 1983) 26, and M. Sakanouye: *Chikyu Kagaku*, **16**: 2 (1983) 7 (all these articles are in Japanese).

[10] Suekichi Kinoshita graduated from Tokyo Imperial University in 1902 and studied in Manchester under Rutherford (1907–1909). He taught and studied radioactivity at his alma mater (1914–1933).

[11] Soddy, F. *The Chemistry of the Radio-Elements*. London: Longmans, Green & Co., 1911.

[12] Yōhachirō Okamoto was serving as a secondary school teacher and a mining expert in the Taiwan Government-General at Taipei (1899–1913) when he discovered hokutōlite in 1906. Taiwan was a territory of Japan at that time. Okamoto did not graduate from any university; he studied mineralogy on his own and under several famous mineralogists. He was a schoolmaster of a primary and secondary private school in Amoy, Fukien Province, China (1913–1935), and he taught mineralogy at Kyushu University, Fukuoka, Japan (1939–1958).

[13] Kaichirō Manabe, a physician, was Professor in the Medical School of Tokyo Imperial University (1926–1937).

[14] Den-ichirō Ishitani, a physicist, graduated from Tokyo Imperial University in 1905 and taught physics first at the Peer's School, then at the First High School (1919–1941), which, after World War II, became the College of General Education, University of Tokyo.

[15] Risaku Ishizu was a technical expert in pharmaceutical sciences at the Tokyo Institute of Hygienic Research, a government institute (1908–1917).

[16] Soddy's paper, *J. Chem. Soc.* **115** (1919) 1–26, was abstracted by Iimori in *J. Chem. Soc. Tokyo* **40** (1919) 536–540.

[17] Iimori published four articles concerning the history of radiochemistry in Japan: (a) 'History of Radioactivity Measurement', *Kagaku-No-Ryōiki* **13** (1959) 689–696; (b) 'Radioactivity and Radiochemistry', *Kagaku-To-Kōgyō* **13** (1960) 510–512; (c) 'The Development of Radiochemistry and My Recollections', *The Tenth Anniversary Volume of the Japan Society of Analytical Chemistry*, Tokyo: 1963, pp. 121–126; and (d) 'My Recollections of Chemical Analysis and Other Subjects in the Past', *Bunseki* (1975) 398. Article (b) is short, and articles (a), (c) and (d) overlap considerably. All these articles are in Japanese.

[18] Shibata, Y. and Kimura, K. 'The Chemical Investigation of Japanese Minerals Containing Rarer Elements. Part I. Analyses of Fergusonite, Naegite, and Monazite, of

Naegi, Mino Province', *J. Chem. Soc. Japan* **42** (1921) 1 (in Japanese). No radioactivity measurements were reported for these minerals.

[19] Eduard Suess (1831–1914), an Austrian geologist, proposed a hypothesis of juvenile or magmatic water to explain the origin of the Karlsbad (now called Karlovy Vary) hot spring in Bohemia: Hot water evolved from the magma deep in the earth comes up through the fissures of rocks to the earth's surface and becomes a hot spring, which may be different in chemical composition from the rain water which had permeated into the earth's crust and come up again.

[20] Masayoshi Ishibashi graduated from Kyoto Imperial University in 1921 and studied under Hevesy (1930–1932) at Freiburg University, Germany. Ishibashi taught analytical chemistry at Kyoto Imperial University (1924–1959) and studied polarography and the abundance of trace elements in sea water.

[21] Yosio Nisina, an atomic physicist, worked as a chief researcher at The Institute for Physical and Chemical Research, Tokyo. In 1928, together with O. Klein, he proposed a formula for the Compton scattering of X-rays – Klein-Nisina's formula. The discovery of U-237 was reported by Y. Nisina, T. Yasaki, H. Ezoe, K. Kimura, and M. Ikawa in *Phys. Rev.* **57** (1940) 1182.

[22] Results of studies on the damages caused by the atomic bombs and on the radioactive deposits in Hiroshima and Nagasaki are reported in the survey report, 'Genshi Bakudan Saigai Chōsa Hōkokushū', 2 vols. (1953), published by the Japan Society for the Promotion of Science (in Japanese).

[23] K. Kimura *et al.*, *Bunseki Kagaku* (Japan Analyst) **3** (1950) 333. Reports on the 'Bikini ashes' were compiled and published in English as 'Researches on the Effects and Influences of Nuclear Bomb Explosions' (1956) by the Japan Society for the Promotion of Science.

[24] Article (b) of note 17.

# PART V

# CHEMISTRY TEACHING AND COMMERCE

A. D. CRUICKSHANK

# 8. SODDY AT OXFORD*

*In the inter-war years Soddy held the Dr. Lee's Professorship of Chemistry at Oxford University. During this period it appeared that Soddy had abandoned scientific research in order to indulge in ill-informed sociological and economic theorising. Insofar as Soddy did become scientifically unproductive, this paper investigates the reasons for his failure to maintain in the inter-war years the high standard of research he had achieved prior to World War I.*

Frederick Soddy's productive pure research ended with the outbreak of World War I. Before that time Soddy was internationally acknowledged as a great scientist. He had, with Rutherford, produced the atomic disintegration theory and, in association with Ramsay, had proved it experimentally. He had been one of the first men to elucidate the nature of isotopes, and it was he who gave them their name. After the war he was dismissed as a man who had forsaken his science to propound wild theories of monetary reform.

The war had a profound effect upon Soddy. Of his many reactions to it not least was his horror at the mobilization of science, and particularly chemistry, to what were, he felt, evil ends. His partial intellectual shift from pure science to the problems of the economy and society had its roots in the effect of the war upon the conduct of science and its impact on the national economy. Whether this shift was also in part due to Soddy's realization of the changing nature of his discipline is uncertain, though he did point out (albeit implicitly) that chemistry had made its innovative contribution to the study of the atom, and physics was to assume the dominant role.

Whatever the reasons behind it, this change in intellectual direction coincided with Soddy's return to Oxford in 1919, and it is with his time at Oxford that this paper is chiefly concerned.

## 1. The Undergraduate

Soddy was born in Eastbourne on 2 September 1877. His mother died while he was still very young, and the rather turbulent upbringing that

*George B. Kauffman (ed.), Frederick Soddy (1877–1956)*, 157–170.

resulted contributed, in Soddy's view, to his introverted and sometimes difficult character. During his early education at Eastbourne College under the guidance of an Oxford science graduate, he showed a keen interest in chemistry and a considerable flair for experimental work, and after a brief period at the University College of Wales, Aberystwyth, he went up to Merton College Oxford, on an Open Science Postmastership – despite an initial failure in Latin at responsions, which made his arrival a term late.[1] He graduated with First Class Honours in the School of Natural Sciences in 1898, but remained in Oxford until the spring of 1900 to engage with a contemporary in research in synthetic organic chemistry. He later admitted that this year was wasted.[2]

In 1899 Soddy applied for the post of Professor of Chemistry at the University of Toronto, and followed up his application by making an uninvited visit to the city.[3] Soddy stood little chance of success: he was, after all, not twenty-three years old, with no teaching or administrative experience and only the merest smattering of research training. Soddy did not get the job. In any event, shortly after his arrival in Toronto he read in a newspaper that the Principal of the University had declared, at the farewell dinner for the departing professor, that the only good chemistry was German chemistry – a view which was also very prevalent in England at the time – and Soddy was able in later years to attribute his failure to this Germanophile attitude.[4] But lured by the new and magnificently equipped MacDonald laboratories at McGill University, Soddy determined to make his way home via Montreal. Fortunately, his arrival there coincided with an attempt by the chemistry department to find suitable English chemists as demonstrators, and he was taken on the staff. His subsequent work with Rutherford on the atomic disintegration theory, his return to London in 1903 to work with Ramsay, and his Nobel Prize-winning research into isotopes and the group displacement law while a lecturer at Glasgow University, are now an important, if previously somewhat underemphasized, part of the history of science.[5]

## 2. Soddy's Return to Oxford

During World War I Soddy was professor of chemistry at Aberdeen. In 1919 he left to take up a chair at Oxford showered with good wishes from both students and staff:[6] but if his old colleagues were not happy to see him go, some of his new colleagues were equally unhappy to see him arrive. The Dr Lee's Chair of Chemistry which Soddy occupied was a new chair, created from the Dr Lee's Readership, and it is said that the electors were divided about the merits of two candidates, Harold Hartley and Neville Sidgwick, both already Oxford dons, and that being unable to agree, they accepted the suggestion that a third person, a distinguished chemist and a graduate of the university, should be appointed. The story may well be apocryphal, but it is consistent with the chilly reception Soddy received from some members of the faculty. As one source says 'My tutor, N. V. Sidgwick, was at little pains to conceal his animosity towards the Professor'.[7] Sidgwick had good reason to be disgruntled; while Soddy had previously applied unsuccessfully for a fellowship at St John's College Oxford, and the Waynflete Chair of Chemistry at Oxford, not to mention the Chair of Chemistry at Aberystwyth, he had not applied for the Dr Lee's Professorship.

For almost the whole of his tenure of the chair Soddy was not a member of the Board of the Faculty, an extraordinary circumstance which was apparently the combined result of two causes. Firstly, at very nearly the same time that Soddy took up his own appointment, Lindemann was appointed professor-physics. Unlike Soddy, Lindemann had not been educated at Oxford and did not qualify for the Oxford MA degree required as a member of the Faculty, so he was duly given one by decree. Soddy, who was eligible for the degree, had not received one because he had never paid the necessary fee, but on hearing that Lindemann had been *given* one, he demanded that the same concession be made to himself, and he refused for many years to buy one. Thus one of the professors of chemistry was not a member of the Faculty, and could neither sit on committees nor attend Faculty meetings. This ludicrous situation was resolved only when a colleague eventually talked him into parting with the money.[8] Secondly, even when Soddy had become a member of the Faculty, he was excluded by certain votes from sitting on the Board. As a colleague later remarked, 'It is inconceivable that in any other university the Professor in charge of one of the biggest teaching departments could be treated in such a way'.[9]

When Soddy arrived at the University, Perkin, the Waynflete Professor (of organic chemistry), had already moved into the new Dyson Perrins laboratory, and Soddy found himself in an old, ill-kept, ill-furnished, and, so far as research was concerned, ill-equipped laboratory, known, appropriately, as the Old Chemistry Department. He was charged with instruction in both physical and inorganic chemistry, and was obliged to lecture in at least two of the three university terms, for at least six weeks per term. The initial stipend was £900 per annum.[10]

## 3. Soddy's Scientific Work

The scientific community expected that Soddy would become the leader of a strong British school of radiochemistry centred at Oxford,[11] but in the event no such thing happened. After his arrival he contributed nothing more to the rapidly progressing study of the atom, and no eminent colleagues joined him. In his later years he complained bitterly that radiochemistry had been taken over by the mathematicians and turned into the unnecessarily complicated subject of atomic physics; yet it might be argued that the radiochemists, as exemplified by Soddy, did not put up much of a fight. Soddy was a practical chemist of the old school, and was not able to adapt his approach to include what were essentially theoretical concepts. His own theories were drawn from experimental observation and then rigorously tested by more experiments, and he was deeply suspicious of hypotheses which could not be tested in such a way.

> ... the wretched layman is deluged with a, to him, utterly meaningless farago of negatrons and positrons, neutrons, protons, nuclei, nucleons, neutrons, mesons, and so-ons, and what-nots before being allowed to study the seminal subject of radioactivity at all ... to make it appear to have originated in the corkscrew brains of the mathematical physicists its very name 'radioactivity' has been changed to nuclear physics. Why not nuclear chemistry?[12]

However, the extent to which his detractors claimed that he had given up radiochemistry was exaggerated. In quantity, research continued in his laboratory much as it had done throughout his career, but the work was not significant. Soddy was secretive about his work (even to the extent of making no further publications), claiming that his research on isotopes had been pirated and that he did not want it to happen again, but it is doubtful in any case whether anyone would have been interested in his results.

He spent a large part of his time perfecting a method for extracting thorium from Travancore monazite sand, presumably with commercial applications in mind, for he began enquiries about the trade position in

1922.[13] After his retirement Soddy was approached by a Glasgow company who were initially keen to take up his process, but could see no market for the product and eventually decided against it, much to Soddy's disappointment.[14] Soddy also studied other radioactive minerals for their constituents, and carried out a large amount of fruitless work on the actinide series of elements.

This run-of-the-mill research was in part due to Soddy's conviction that the important work on the atom had finished with the discovery of isotopes and the group displacement law, and all that was left was a mopping-up operation.[15] As a scientist Soddy may have been in error, but as a radiochemist he was in some senses right. Concerned mainly with the purification, extraction, and analysis of radioactive elements and their compounds, radiochemistry could not be in the forefront of the investigation of the nucleus, and physics was to assume the dominant role.

### 4. Soddy's Activities within the University

The arrogant assumption that he had done all the important work and exhausted the conceptual possibilities of his subject appears typical of Soddy at Oxford. Rather in contrast to the picture of him gleaned from his time at Glasgow and Aberdeen, at Oxford he appears to have been both overbearing and dogmatic. If he had any tendency towards immodesty before 1921, the award of the Nobel Prize in that year could only have exaggerated it. Indeed, it is said that the prize gave him an inflated sense of his own importance, and a conviction that in all things a man of his intellect must be right.[16]

Soddy was deeply suspicious of Oxford's administrative and educational policies even before he arrived:

> . . . the curricula of the ancient universities accumulate rather than evolve . . . the whole system of bursaries and regulations for degrees is to perpetuate and bolster up a museum of ancient learning, and the system of finances to divert to its support the resources needed for living subjects. . . . The result is that the ancient universities become not by any means the quiet sanctuaries of ancient learning, which would be relatively harmless, but the active agents in perpetuating in power a type of man who is hopelessly out of tune with his environment.[17]

And he was fully aware of what he was taking on when he accepted the post of Dr Lee's Professor. It is, however, possible that Soddy both overestimated his own strength and underestimated the magnitude of the inertia for which the ancient universities were famed.

The distribution of government money within the University was not, Soddy felt, fair to science. The funds were channelled through the standard financial and administrative councils and committees '... on which the old vested interests reign supreme',[18] not only controlling, but also

> striving by every means to emulate the expansion of science in their staffs, if not in their students, in order to preserve their majority on the councils. Research to such is at most a pious aspiration whereas the running of the universities and making the accounts balance is their business.[19]

The ancient universities were not able, or not willing, to help science:

> ... their resources go to maintain museums of learning and are appropriated to the old subjects, to the Church, to the classics, and to mediaeval and drawing-room studies.[20]

While the ancient universities were, in their ponderous manner, already beginning to accommodate the inevitable emergence of science, Soddy was determined to hurry things on a little.

The laboratory provided for Soddy was far from ideal, and he was distressed not only by this but also because he found that a professor of chemistry at Oxford did not command the same amount of power as he did at a Scottish university. For example, the practical teaching of undergraduates was undertaken in the college laboratories, and any suggestion that Soddy should have control over it was strongly resisted. When, shortly after he was appointed, a committee was set up by the Hebdomadal Council to consider the future of chemistry in the University, Soddy submitted a minority report in which he pleaded for the replacement of the college laboratories by properly built and fully-equipped university laboratories, and made other proposals, most notably concerned with undergraduate tutorial arrangements. While all his suggestions were at the time rejected both by the committee and by Council, nearly all of them were later put into practice. No attempt was made to build a university laboratory for physical chemistry until he had retired, at which time plans were immediately put in hand for the purpose.[21]

There were various reasons for the poor reception given to Soddy's proposals. His arrogance, together with a complete lack of amicability or willingness to compromise, antagonized the very people he had to impress. Soddy made no attempt to endear himself to his colleagues, and under these circumstances sheer force of argument was not enough. In any case, some of Soddy's criticisms were not supported by either students or dons,[22] who considered that the practical and tutorial teaching systems were both adequate and efficient. Soddy's attempted reforms had the appearance, true or false, of a drive for personal influence and power, and they did not succeed. A confrontation with the university authorities over a departmental appointment which had been made without consulting him only served to illustrate the weakness of his position and the extent of his frustration.[23]

Eventually Soddy was able to improve his working conditions, for after he had been at Oxford about eight years he was given some twelve or thirteen thousand pounds with which to carry out the entire reconditioning of the Old Chemistry Department. This, he said, 'absorbed my entire energy for some years',[24] which is not as surprising as it might appear because Soddy typically acted as his own architect, builder, and engineer. While he succeeded magnificently in carrying out the reconditioning, he was frustrated in his attempt to build a much-needed extension by lack of funds and lack of university support. Again, Soddy was his own worst enemy, for, so the story goes, when Thomas Lodge, who had been appointed by the Hebdomadal Council to plan the layout of the science area as a whole, visited him to see the models of his proposed new extension, Soddy became enraged, and dismissed Lodge with the words 'I don't need your help or that of any other * * * * architect'.[25]

Soddy's intransigence was not one of his most appealing features, yet in his dealings with the University it appears to have been the most characteristic. Such an attitude was ill-suited to a university such as Oxford, where compromise was essential to overcome the inconsistencies inherent in a system whereby professors were in charge of the laboratories (such as they were), while the faculties were largely in the hands of college tutors.

Soddy was disturbed that, in contrast to his time at the Scottish universities, at Oxford he was unable to establish any sort of relationship with the students. The undergraduates were under the immediate influence of their tutors, and few graduates attached themselves to his laboratory for research training; his own research work is considered to have suffered as a result.[26] This is perhaps an indication not only of the effect of his aloofness and reticence on other people, but also of the low standing this recent Nobel Prizewinner, and the subject of his research, held in the scientific community.

## 5. Soddy's Activities Outside the University

Soddy found plenty to do which did not involve radiochemistry at all. The war forced him to the unpalatable conclusion that science was proving 'as much a curse as a blessing to mankind',[27] and he spent a good deal of his time wondering how this 'good tree' of science could bring forth such evil fruit. By the end of the war he had decided, along with many other scientists, that the evil fruit had been grafted on to the good tree by the agents of State and Economy. He assumed that science was inherently not merely neutral, but positively beneficial, at least in the long run, and that for its maleficent effects society, not science, was to blame. The damaging convulsions which society was undergoing, directly or indirectly as a result of scientific progress, were, Soddy felt, due to the obsolete nature of the social system, which was incapable of maturing even as it was dragged along in the wake of scientific progress:

> [It is] as unable to imagine anything not hallowed by tradition as it would be to grasp the significance of a new dimension. A medley of archaism and mediaevalism, masquerading in a modern garb that does not fit, a patchwork system, wherein lingering relics of obscurantism and feudalism co-exist with the horros of a modern industrialism and an antiquated imperialism that have been made drunk with the new wine of science.[28]

Soddy's thinking on this subject followed fairly closely that of Arnold Toynbee.[29]

While Soddy gave much thought to the relationship between science and society he was never associated with either of the radical or reformist factions which formed the popular scientific front of the inter-war period.[30] Indeed, while a suggestion that there should exist a society for the social relations of science was welcomed by such men as Julian Huxley, Wells, and Tizard, Soddy contemptuously dismissed the proposed new institution as just another debating society – like the Royal Society so encumbered by its chains of office as to be totally ineffective.[31]

The Royal Society was another of Soddy's major targets for reform. In 1935, together with some thirty other Fellows, he started a well-publicized revolt against the administration, arguing that the Society did not properly represent science to the nation, and was not subject to proper democratic control. Worthy people were excluded from membership:

It is a serious charge, but I say quite definitely, that many a scientist that deserves election to the Royal Society has been kept out because he does not know the right people.[32]

While Soddy and his colleagues were not immediately successful in their attempt to undermine the Society's administration, he remained typically unperturbed and confident:

The battle for reorganization has started, we shall not cease to fight till we have turned the Royal Society into an organization that shall be for the scientist what the BMA is for the doctor.[33]

The representative scientific societies, the British Association for the Advancement of Science and the Royal Society, were indeed becoming increasingly establishment-oriented, and with the new-found social awareness among scientists it was inevitable that a new, activist organization should be formed. This was the National Union of Scientific Workers, and from 1920 onwards it commented outspokenly on matters of science policy, it views influenced by the rising group of radical scientists which in this case included Soddy. However, by 1927 the National Union had begun to revert to a professional organization; the name was changed to the Association of Scientific Workers, and the Association followed a route towards a simple trade union for technicians and industrial scientists.[34]

## 6. Soddy and Monetary Reform

The years during and after the war were years of economic restraint for the Government, and grinding poverty for many of the people, and it was at this time that Soddy commenced a systematic study of economics as a subject (though his interest may be traced back to his youth). He felt that whereas science advanced further and further *away* from the problems of life and everyday existence, economics advanced *towards* them – and he wanted to find a cure for the country's economic ills.

Soddy considered the Government demands for economy lidicrous when science was capable of creating vast amounts of wealth.

It is preposterous that in an age of actual and potential plenty, rendered possible by scientific development, the only cry should be economy.[35]

It was, however, manifestly obvious that the wealth which science was in the process of creating was not filtering properly through society, and Soddy deduced that the fault lay with the means of distribution, that is, with the monetary system. In his attempt to correct the situation he formulated a version of Douglas's theory of social credit,[36] and sought to apply the laws of conservation to economic analysis. Money, he argued, was not itself wealth, but only the means by which wealth could be obtained from the rest of society (in the form of goods and services). He therefore identified a new category of wealth, which he termed 'virtual wealth', as the amount of money the community possessed and thus as the amount of real wealth which it was willing to forego. He suggested that 'The physically impossible [in terms of the laws of conservation] quality which in our times has been grafted upon [money] is that it should bear interest',[37] and he condemned the banks for using interest to increase the amount of money in circulation without similarly increasing the amount of available real wealth, thereby debasing the currency. Soddy suggested that the banks be nationalized, and a national currency instituted.

Though eminent, Soddy was only one of a large group of monetary reformers (or 'currency cranks' as they were less politely known), all of whom were more or less ignored by academics and politicians alike. There was, nevertheless, some popular interest in his work, and he was frequently invited to lecture on his theories. Indeed, even when he was speaking on a completely different topic, economics usually managed to intrude. For example, he was asked to give a lecture on 'Science and religion' to the Oxford University Student Christian Movement. Two students were later overheard discussing it.

> 'He didn't say much about religion, but when he got on to economics he was very interesting.'
> 'Oh. That's his subject isn't it?'
> 'Yes, I rather think it is'.[38]

And at the time, it was.

## Soddy's Retirement: Conclusion

During the later part of his career at Oxford Soddy became interested in two other subjects, mathematics and mechanisms. Of the former, he continued his mathematical work until late in his life, and he gave instructions for it to be carefully preserved after his death. He published material on harmonic series and the problems of 'kissing circles'.[39] Of the latter, many of his inventions were patented but none was developed for further use, although a machine which assisted in the identification of spectral lines was used in the laboratory and another, a reversing centrifugal gear, was shown at the Eighth International Exhibition of Inventions in London in 1932.[40]

In 1936 Soddy's wife, whom he had married in 1908, died, and he lost the strength and support which had assisted him through his difficult years at Oxford. He promptly resigned the professorship, though in explaining his decision the death of his wife remained unmentioned: '... when I resigned it was with the intention of making way for a younger man rather than giving up my experimental work'.[41] Soddy obtained rooms in the Old Museum House (now demolished) in which to work, and in the few years before World War II was frequently seen entering and leaving his makeshift laboratory.[42]

Soddy had few friends in Oxford, and on the outbreak of World War II he moved first to Enstone, some twenty miles north of Oxford, and finally to Brighton where he lived a despondent and somewhat embittered life until his death in 1956. While Soddy viewed the progress of science with growing distrust (which reached its peak after the explosions at Hiroshima and Nagasaki), he was bitter because he had lived to see himself (along with other radiochemists) virtually written out of the history of science in favour of the later work of physicists such as those at the Cavendish Laboratory. His work was made to appear slight and unimportant, and this absolutely infuriated him.

> ... the absolute travesty of scientific history which the little gang of mutual back-scratchers dominating British science has invented is to anyone who has lived through the period well-nigh unbelievable. If you had asked me whether 50 years after the first discovery and proof of a natural transmutation of the elements, this key work which introduced the atomic age could be so suppressed in favour of the bits and pieces of a bragging school of advertisers as to be completely unknown to the modern generation, I should have thought that quite impossible ...
>
> The step almost always made to appear as the start, instead of the 1902–04 discoveries, is when Rutherford in 1919 (though ... this explanation was not forthcoming till many years afterwards) transmuted the atom of N into one of O by bombarding it with alpha-particles.[43]

Soddy also aimed his caustic remarks at a specific target.

I'm sick to death of the efforts of the Cambridge clique to try and falsify the whole history of the subject so as to concentrate the limelight on the Cavendish laboratory.[44]

After his death many people wrote complimentary pieces about Soddy, emphasizing his scientific achievements, glossing over his time at Oxford, and treating his (major) excursions into economic and social problems as regrettable aberrations. Others were less complimentary:

The distinguished scientist who is getting a bit tired of science and who seeks to give the world the benefit of his views on social questions is, unfortunately, a frequently encountered phenomenon, and Soddy was the first of these....

Soddy's writings must bear a considerable responsibility for creating the philistinism and lack of any orientation, philosophical or cultural, of subsequent generations of scientists.[45]

These were, however, gratuitous criticisms. Soddy never tired of science. Having previously single-mindedly pursued his scientific research, during and after World War I, Soddy simply diversified putting his mind to economic and social problems and their relation to scientific progress. He continued his radiochemical research (he searched in vain for what is now known as the actinide series),[46] but it was no longer his inspiration.

Soddy's task was not that of a philosophical or cultural leader; he was a researcher, a problem-solver. But he was never one to adopt a position of sanguine isolationism, and wherever he saw a situation which he thought needed to be changed – be it in the arrangements for teaching Oxford chemistry undergraduates, or in the monetary system of Great Britain – he strove to bring that change about. Above all he was concerned to ensure that any further contributions from science to civilization would not be abused by the very society they were intended to bless. That he was not immediately successful can perhaps be ascribed to his abrasive character and behaviour as much as to the unorthodoxy of his message. Unfortunately, the poor reception which the latter received, ranging from contemptuous dismissal to outright hostility, only served to exacerbate the former; and both were compounded by his institutional isolation within the University of Oxford.

*Nuclear Engineering International*
*Quadrant House*
*Sutton*
*Surrey SM2 5AS*
*England*

## Notes

* Reprinted from *The British Journal for the History of Science* **12**: 42 (November 1979): 277–88.

[1] General biographical details have been obtained from the following sources: F. Soddy, 'Extracts from Royal Society, "Personal Records of F. Soddy"', Soddy Collection, Bodleian Library, Oxford, Catalogue Number 2 (Hereafter SCB 2); F. Soddy, 'Nobel prizeman in chemistry 1921 (awarded 1922)', 7 July 1950, (SCB 2); A. Fleck, 'Frederick Soddy', *Biographical memoirs of Fellows of the Royal Society*, 1957, **3**, 203–13; A. Fleck, 'Frederick Soddy' (Obituary), *Nature*, 1956, **178**, 893; F. Paneth, 'A tribute to F. Soddy', *Nature*, 1957, **180**, 1085–7; E. J. Bowen, 'Note on Frederick Soddy', 29 January 1974 (SCB 1); F. M. Brewer, 'Frederick Soddy', in M. Howorth (ed.), *An appreciation of Professor Frederick Soddy*, London, 1956, pp. 12–15 (SCB 28).

[2] F. Soddy, 'Extracts', *op. cit.* (1).

[3] Testimonials from Odling, Harcourt, and Ramsay are in SCB 5. See also F. Soddy, 'Extracts', *op. cit.* (1); and 'Just fifty years on', lecture given to the Institute of Atomic Information for the Layman, 28 February 1953 (SCB 26).

[4] *Ibid.*

[5] See, for example, F. Soddy, 'Résumé of Original Investigations, 1900–1913, of Frederick Soddy' (SCB 3); 'The Joint Researches of Sir William Ramsay and Frederick Soddy', March 1946 (SCB 2); 'The Origins of the Conception of Isotopes', Nobel Lecture delivered at Stockholm, 12 December 1922 (Stockholm, 1923). For biographical and scientific details (and further bibliography) see T. J. Trenn, 'Soddy, Frederick', in C. C. Gillispie (ed.), *Dictionary of Scientific Biography*, New York, 1976, xiii, 504–9. I have written my own commentary on this work: A. D. Cruickshank, 'Soddy, Science, and Society', University of Oxford BA (Chemistry) Part II dissertation, 1976.

[6] 'Professor Soddy', *The Crucible*, 1919, **2**, 17–19 (SCB 15); W. T. H. Williamson, 'Soddy's War-Work at Aberdeen' (SCB 32); for a recent study see Kenneth R. Page, 'Frederick Soddy: The Aberdeen Interlude', *Aberdeen University Review*, 1979, **40**, 127–48.

[7] F. M. Brewer, 'Statement', undated typescript in SCB 28.

[8] E. J. Bowen, private communication.

[9] F. M. Brewer, 'Frederick Soddy', *op. cit.* (1).

[10] *Oxford University Gazette*, 1919, **49**, 283.

[11] F. Paneth, *op. cit.* (1), 'Tribute'.

[12] F. Soddy, 'The Wider Aspects of the Discovery of Atomic Disintegration – Contrasting Experimental Facts with the Mathematical Theories', address to the Fourth conference of Nobel prizewinners, 30 June 1954, (London, 1954) (SCB 44).

[13] F. Soddy to High Commissioner for India, 19 August 1937 (SCB 20).

[14] Correspondence between F. Soddy and Stanford and Co Ltd is in SCB 21.

[15] E. J. Bowen, private communication.

[16] *Ibid.*

[17] F. Soddy, 'Science and the state', presidential address to Aberdeen University Scientific Society, in F. Soddy, *Science and Life*, London, 1920, chapter IV.

[18] F. Soddy, 'The Public Support of Scientific Work', address to the National Union of Scientific Workers, *Scientific Worker*, 1920, **1**, 15–20 (SCB 44).

[19] *Ibid.*
[20] *Ibid.*
[21] F. M. Brewer, 'Statement', *op. cit.* (7).
[22] E. J. Bowen, private communication.
[23] F. Soddy, 'Petition – To the King's Most Excellent Majesty in Council – The humble petition of Frederick Soddy', 19 September 1927 (SCB 18). 'Reply of the University of Oxford to the Petition dated 19 September 1927 – To the King's Most Excellent Majesty in Council – The humble petition of Frederick Soddy', 21 March 1928 (SCB 18).
[24] F. Soddy, 'Extracts', *op. cit.* (1).
[25] Quoted in F. M. Brewer, 'Frederick Soddy', *op. cit.* (1), and 'Statement', *op. cit.* (7).
[26] E. J. Bowen, 'Note', *op. cit.* (1).
[27] F. Soddy, 'Extracts', *op. cit.* (1).
[28] F. Soddy, 'Scientific and Sociological – The Social Abuse of Science', lecture delivered at Aberdeen, 1 November 1918 (SCB 114).
[29] A. J. Toynbee, *A Study of History*, 18 vols., Oxford, 1962.
[30] P. G. Werskey, 'British Scientists and Outsider Politics, 1931–45', *Science Studies*, 1971, **1**, 67–83.
[31] F. Soddy, 'Extracts', *op. cit.* (1).
[32] *The daily telegraph*, 3 December 1935. Material pertaining to this conflict is in SCB 10.
[33] *Ibid.*
[34] H. Rose and S. Rose, *Science and Society*, Harmondsworth, 1971, 52–4.
[35] Quoted in 'Professor Soddy against the Government – Labour Instincts Are Right', *Forward*, 26 September 1931, p. 7.
[36] C. H. Douglas, *Economic Democracy: Credit Power and Democracy*, London, 1920. Soddy's economic thinking is more fully detailed in the article by T. J. Trenn, chapter 10.
[37] F. Soddy, *The Inversion of Science, and a Scheme of Scientific Reformation*, London, 1924, p. 10.
[38] F. M. Brewer, 'Statement', *op. cit.* (7).
[39] F. Soddy, 'Summation of Infinite Harmonic Series', *Proceedings of the Royal Society*, 1941–2, **179A**, 377–86; 'The Three Infinite Harmonic Series and Their Sums', *ibid.*, 1943–4, **182A**, 113–29; 'The Kiss Precise', *Nature*, 1936, **137**, 1021; 'The Hexlet', *ibid.*, p. 958. Soddy's mathematical notebooks and the card on which he wrote his instructions to his executors are in SCB 46–9.
[40] *Nature*, 1932, **130**, 573–4.
[41] F. Soddy to N. Hinshelwood, 6 February 1937 (SCB 236), and other material in SCB 236.
[42] *Ibid.*
[43] F. Soddy to M. Howorth, 22 February 1955 (SCB 29).
[44] F. Soddy to M. Howorth, 25 December 1952 (SCB 29).
[45] C. Clark, 'The Aberration of Genius', *The tablet*, 13 October 1956, pp. 298–300 (SCB 45).
[46] F. M. Brewer, 'Statement', *op. cit.* (7).

MICHAEL I. FREEDMAN

# 9. FREDERICK SODDY AND THE PRACTICAL SIGNIFICANCE OF RADIOACTIVE MATTER*

*Frederick Soddy, best known for his purely scientific work in radiochemistry, also took an interest in the practical significance of his science. He was concerned with the immediate issue of the commercial availability of radioactive substances. He also considered the long-term issue of the utilization of the energy of radioactive substances. A probable factor in the development of this interest was Soddy's relationship with the industrial chemist, George Beilby. For Soddy, radioactive matter was important for its bearing upon human affairs.*

It is for his scientific achievements that we best remember Frederick Soddy – first as the young chemist working with Ernest Rutherford and with William Ramsay in elaborating the disintegration theory of radioactive change, and then as the mature chemist, heading a research programme of his own at the University of Glasgow, a programme which culminated in his formulation of the concept of isotopes in the years before the First World War.[1] Yet there was another side to Soddy's early scientific career: beyond his profound concern with the purely theoretical and experimental aspects of radioactivity research, there was a serious interest in what might be called the practical significance of his science. By practical, I mean those aspects capable of being put to use for the immediate or potential future benefit of man. In this paper I wish to elucidate the nature of this concern during Soddy's career, focusing particularly upon his years at the University of Glasgow, 1904–1914.

For Soddy, the practical significance of radioactive matter lay primarily in two domains. It was important first in the short run, as a tool for medicine and also, owing to the rising demand for radium by physicians and others, and the concurrent shortage of supply, as a commodity for commerce as well – a valuable product of manufacture. And in the long run, radioactive matter was of great potential significance as a fundamentally new source of energy for industrial society. Both of these domains were important to Soddy, and I will examine each in turn, and suggest a possible source of his interest.

The immediate issue of commercial availability was of importance to Soddy from 1903 on. Soddy being a researcher, his interest had its

*George B. Kauffman (ed.), Frederick Soddy (1877–1956)*, 171–176.

origins in the nature of his early scientific experiments, in particular, his work with Ramsay, during which the entry of a new radium manufacturer into the market had suddenly made accessible to them a hitherto unavailable amount of radium salt at a low cost, and made possible their study of the production of helium from radium. Recognizing its significance for scientific research, Soddy from then on began to call for the increased production of radium. But he was also concerned with its availability for medical purposes. Soddy himself was an innovator of a therapeutic application of radon gas for the treatment of tuberculosis. More particularly, while at Glasgow Soddy became vocal about the shortage of radium supplies for Britain, and contributed to the movement to establish radium resources for the West of Scotland.[2]

Perhaps the most striking example of the seriousness with which he took this practical issue of commercial availability, was the development of his patent of 1910. It was early in 1909 that Soddy saw a solution to the radium shortage in a newly discovered radioelement mesothorium; this substance, with its decay products, had the same penetrating radiations as radium, but could be manufactured from a cheaper and more plentiful source, thorium ore, widely used in the gas lighting industry. Soddy called upon thorium manufacturers to begin producing this radium substitute to help satisfy the rising demand for radium.[3] By 1910, however, it was clear that the knowledge necessary for mesothorium production – specifically, the characterization of its chemical properties – was being kept secret by its discoverer, Otto Hahn, then associated with a German thorium manufacturer, Knoefler and Company; this was apparently because such knowledge was of potential commercial value. Responding to this secrecy, and recognizing the importance of this chemical knowledge for commercial as well as for his own laboratory use, Soddy sought to determine it for himself, embarking on a remarkable project.

By November 1910, Soddy had succeeded in establishing the chemical similarity of this radioelement to barium, and with this knowledge he registered preliminary specifications for a patent on mesothorium production. Subsequent experiments led to his realization by mid-December that mesothorium was in fact chemically identical to radium. The purely scientific significance of this finding is, of course, well known to historians of science: Soddy's recognition of this chemical identity marked the starting point for a totally new research programme on the nature of radioactive isotopes, for which he later received the Nobel

Prize. What is not generally known is that this knowledge was also of great practical significance, for it provided the basis of the final specifications of Soddy's patent. The motivation for Soddy's patent is not certain. Clearly it was not simply for private gain, for Soddy was quick to publish the general principles of mesothorium production for others to read and use. But if the intentions are not clear, the results were. Soddy's patent, approved in 1911, threatened the enterprises of the German firm, Knoefler and Company, which by then had been marketing mesothorium but had not patented Hahn's process. In the end, the firm had to pay Soddy for the patent. This episode, then, provides a clear indication of Soddy's actions and concern with the problem of the commercial availability of radioactive substances.[4]

Soddy's interest in the immediate practical significance of radioactive matter was complemented by his apparently prophetic concern with its long-term implications for society. Of great importance to him was the energy contained within radium and the other radioelements. Again his interest had its origins in the nature of his early scientific experiments; in this case, his work with Rutherford, in which the two were among the first to calculate the tremendous amounts of energy associated with radioactive change. These findings suggested to Soddy that this constituted a new source of energy which might one day be utilized for practical ends, if the means could only be found to release this energy rapidly by a process of artificially accelerating radioactive change.[5]

Especially at Glasgow, Soddy gave wide publicity to this practical implication of radioactivity research. In his address to Glasgow electrical engineers in 1906, in his public University lectures in 1908, and in his numerous writings of the time, Soddy consistently and emphatically portrayed atomic energy as providing the basis for a new material civilization, and a solution to the problem posed by the inevitable depletion of the coal reserves necessary for industrial society.[6] While Soddy was not alone in his belief in the possibility of exploiting atomic energy, he was perhaps the most visionary and widely read of his contemporaries. We may gauge the power of Soddy's writings on this practical significance by the fact that his popular book, *The Interpretation of Radium*, provided the inspiration for a science fiction novel by H. G. Wells, *The World Set Free*, published early in 1914. Wells acknowledged his debt to Soddy, and Soddy in turn regarded this as Wells's greatest novel. Wells portrayed Soddy's prophecy of the utilization of atomic energy in the future – in the novel, this occurred in 1933 – but he also portrayed its subsequent applciation to military purposes, resulting in a global atomic war. Wells's vision in turn became an object of Soddy's concern after the outbreak of the First World War. The prospect of atomic weapons was an awful possibility that Soddy himself came to consider in his addresses of that time. Thus Soddy continued to display a keen appreciation of the practical significance of his science.[7]

Elucidating the sources of Soddy's concern with these wider aspects of radioactivity is a difficult task, especially since Soddy was not – properly speaking – an applied scientist. Of course it was the specific nature of his early scientific work that initially aroused his concern. But it is probable that another important factor in the development of Soddy's interest during his years at Glasgow, was his relationship with his friend and father-in-law, the noted industrial chemist, George Beilby, then general manager of the Cassel Cyanide Company of Glasgow. During these years, Beilby involved Soddy in some of the practical scientific problems of his firm, and in turn provided aid in radioactivity research. We may note in particular that it was Beilby's firm that provided support for Soddy's work on the radium-substitute. Moreover, Beilby's other major position, as a leading British authority on the issue of fuel reserves and on the problem of coal depletion, undoubtedly helped promote Soddy's interest in the issue of energy resources. Lastly, the general concern with social issues in the Beilby household, in which Soddy took part, may have given him a fuller perspective on the relations of science and society.[8]

For Frederick Soddy radioactivity research was far more than simply a means for obtaining a better understanding of the nature of matter in its own right. Radioactive substances also had a practical significance – in the short run, as a tool for medical therapy and as an object of commercial value, and in the long run as a source of energy for industrial society and as a possible means of man's destruction. Radioactive matter had enormous bearing upon human affairs.

*P.O. Box 30459*
*Baltimore, MD 21270, U.S.A.*

## Notes

* Reprinted from *The British Journal for the History of Science* **12**: 42 (November 1979): 257–60.

[1] Muriel Howorth, *Pioneer Research on the Atom: the Life Story of Frederick Soddy*, London, 1958. Alfred Romer, *Radiochemistry and the Discovery of Isotopes*, New York, 1970. Frederick Soddy, *Radioactivity and Atomic Theory* (ed. by Thaddeus J. Trenn), London, 1975.

[2] Howorth, *op. cit.* (1), pp. 98–9. Frederick Soddy, 'Possible Future Applications of Radium', *The Times Literary Supplement*, 17 July 1903, pp. 225–6; 'A Method of Applying the Rays of Radium and Thorium to the Treatment of Consumption', *British Medical Journal*, 25 July 1903, **2**, 197–9; 'Radium Resources', *Nature*, 27 November 1913, **92**, 376–7. Glasgow and West of Scotland Radium Committee minute book, University of Glasgow Archives, item 35904.

[3] Frederick Soddy, 'Radiothorium', *Nature*, 4 March 1909, **80**, 12.

[4] Frederick Soddy, Notebooks, Oxford University Archives, Frederick Soddy Papers, items 82 and 83. 'A Process for the Separation of Mesothorium from Monazite Sand and for the Concentration of Mesothorium in Crude Products Obtained from Minerals Containing Thorium', British patent No 25,504; date of application, 3 November 1910; complete specification, 7 April 1911; accepted 17 August 1911. 'The Chemistry of Mesothorium', *Proceedings of the Chemical Society*, 30 December 1910, **26**, 336–7; 'The Chemistry of Mesothorium', *Journal of the Chemical Society*, 1911, **99**, 72–83. 'The Origins of the Conceptions of Isotopes', in *Nobel Lectures, Chemistry*, Vol. i: *1901–1921*, New York, 1966, pp. 381–2. Ernest Rutherford, letter to Bertram Boltwood, 21 October 1911, in Lawrence Badash (ed.), *Rutherford and Boltwood: Letters on Radioactivity*, New Haven, 1969, p. 255. Otto Hahn, letters to Ernest Rutherford, 4 October 1911, 18 October 1911, in the correspondence of Ernest Rutherford, American Institute of Physics, Reel C/H 56–7. Otto Hahn, *A Scientific Biography* (tr. and ed. by Willy Ley), New York, 1966, pp. 40–8; idem, *My Life: The Autobiography of a Scientist*, New York, 1970, pp. 82–5.

[5] Ernest Rutherford and Frederick Soddy, 'Radioactive Change', *Philosophical Magazine*, 1903, **5**, 576–91. Frederick Soddy, 'The Disintegration Theory of Radioactivity', *The Times Literary Supplement*, 26 June 1903, p. 201; 'Possible Future Applications of Radium', *ibid.*, 17 July 1903, pp. 225–6; 'Radium: A Lecture Delivered to the Corps of Royal Engineers', *Professional Papers of the Royal Engineers*, 1904, **29**, paper VIII; *Radioactivity: An Elementary Treatise, from the Standpoint of the Disintegration Theory*, London, 1904.

[6] Frederick Soddy, 'The Internal Energy of the Elements', *Journal of the Institution of Electrical Engineers*, 1906, **37**, 140–7; 'The Evolution of the Elements', *Report of the 76th Meeting of the BAAS*, 1906, pp. 122–31; *The Interpretation of Radium: Being the Substance of Six Free Popular Experimental Lectures Delivered at the University of Glasgow*, London, 1909, Chapter XI; 'The Energy of Radium', *Harpers Monthly Magazine*, December 1909, **120**, 52–9; 'Transmutation, the Vital Problem of the Future', *Scientia*, 1912, **11**, 186–202; *Matter and Energy*, London, 1912, Chapter X; 'The Evolution of Matter', published in the *New York Times*, 28 September 1913, under the title, 'Science to Revolutionize all Existence'; 'Science and Life', *The Candid, a Quarterly Review of Public Affairs*, 1914, **1**, 237–60.

[7] H. G. Wells, *The World Set Free: a Story of Mankind*, London, 1914. Frederick Soddy, Inaugural Lecture at the University of Aberdeen, 16 October 1914, in Howorth, *op. cit.* (1), p. 194; *Science and Life: Aberdeen Addresses*, London, 1920.

[8] H. C. H. C[arpenter], 'Sir George Beilby, 1850–1924', *Proceedings of the Royal Society*

*of London*, 1925, **109A**, i–xvii. Frederick Soddy, 'Report on the Electric Furnace Experiments at [Cassel Cyanide Co's works at] Maryhill [Glasgow, under the auspices of George Beilby]', and 'Specification of the Accumulator Battery Installed ... in the Experimental Shed at the Cassel Cyanide Co's Works, Maryhill, Glasgow', Oxford University Archives, Frederick Soddy Papers, item 103; Frederick Soddy, letter to Ernest Rutherford, 2 April 1912, in the correspondence of Ernest Rutherford, American Institute of Physics, Reel E/S 157. Alexander Fleck, 'Early Work in the Radioactive Elements', *Proceedings of the Chemical Society*, November 1963, p. 330. G. T. Beilby, 'Modern Power Production and its Relation to the Coal Resources of Great Britain', *Nature*, 22 February 1906, **73**, 398–9. Fritz Paneth, letter to Muriel Howorth, 4 April 1957, in Howorth, *op. cit.* (1), pp. 317–18. G. T. Beilby, *Chemistry and Practical Life*, Glasgow, 1914.

# PART VI

# ENERGY AND ECONOMICS

THADDEUS J. TRENN

# 10. THE CENTRAL ROLE OF ENERGY IN SODDY'S HOLISTIC AND CRITICAL APPROACH TO NUCLEAR SCIENCE, ECONOMICS, AND SOCIAL RESPONSIBILITY*

*In 1903 Soddy contributed to the discovery of a new source of energy within atoms. He initially lobbied for the importance of scientific research to achieve artificial transmutation in order to permit society to benefit from atomic energy. But appalled by the abuse during World War I of such scientific achievements as the nitrogen fixation process, he tried to draw attention to the potential misuse of nature's gifts and refrained from such further research that might accelerate the achievement of controlled atomic energy. Rather he sought the reason for such perversion of science and focussed his attention on an obsolete economic system requiring periodic convulsions to stimulate production and involving a method of redistribution that was far from scientific. Energy – not money – was considered to be the real wealth of nations. Between the two world wars Soddy stressed the need to supplement our limited resources of coal and oil to insure an adequate supply of future energy so as to forestall the economic disorder which would accompany the otherwise inevitable energy crisis. Atomic energy could fulfill this social and economic need but only if it were used peacefully, making this the last viable option according to Soddy for the survival of Mankind.*

> The fact remains that, if the supply of energy failed, modern civilization would come to an end as abruptly as does the music of an organ deprived of wind. [But] ... the still unrecognized 'energy problem' ... awaits the future ...[1]
>
> F. SODDY, 1912

> [The human control of atomic energy could] virtually provide anyone who wanted it with a private sun of his own.[2]
>
> F. SODDY, 1915

> The energy available for each individual man is his income.[3]
>
> F. SODDY, 1918

> [The blame for the future plight of civilization] must rest on scientific men, equally with others, for being incapable of accepting the responsibiity for the profound social upheavals

*George B. Kauffman (ed.), Frederick Soddy (1877–1956)*, 179–198.

> which their own work primarily has brought about in human relationships.[4]
>
> F. SODDY, 1953

## 1. Introduction

Frederick Soddy (1877–1956), one of the foremost radiochemists of his day, was awarded the 1921 Nobel Prize in chemistry. Soddy was also among the first of the scientific leaders of his age, along with Blackett (1897–1974), Bernal (1901–71), and others, to become interested in the social implications of their work. In 1950 his colleague Paneth wrote that currently 'there is widespread discussion on the responsibility towards the community of men of science and particularly experts in radioactivity; but a perusal of Prof. Soddy's non-chemical writings of no less than thirty years ago [viz., during and after the first world war] shows how strongly he felt the duty to fight for a better order of things'.[5] Soddy was a complex iconoclast often derided for dealing with what seemed to be disparate interests. What, it could be asked, could monetary reform possibly have to do with radiochemistry? But to make sense of Soddy, the question must rather be formulated in the other direction: What fundamental concern did Soddy have that enabled him to embrace holistically a variety of seemingly diverse activities? The answer can be given in a word: energy. And the way in which Soddy dealt with this issue involved him directly with social economics as well as with the social responsibility of scientists.

Soddy construed energy, especially the potentially unlimited supply stored within the atoms of matter, as nature's special gift to humanity, available only through the mediation of science. It was both our common heritage and a real part of the wealth of nations. The discovery of this new source of energy, the promise of its control, and the frustration of that promise by non-scientific factors form the theme of the present paper. An examination of this theme from Soddy's complex point of view will hopefully serve to correct the fragmented and superficial received view of Soddy which is still all too prevalent today.

## 2. The Need for, and the Discovery of, a New Source of Energy

In his presidential address to the British Association at Bristol in September 1898, Sir William Crookes directed attention to the limitation of world resources. The solution he proposed lay in the artificial production of nitrate in order to stimulate the production of wheat. 'The fixation of nitrogen is vital to the progress of civilized humanity. Other discoveries', said Crookes, merely 'minister to our increased intellectual comfort, ... [and] hasten the acquisition of wealth ...' Crookes warned of the peril for 'the great Caucasian race ... unless we can class the fixation of nitrogen among certainties to come ...'.[6] Although he was referring specifically to the potential food shortage awaiting civilization without new chemical discoveries, Crookes concluded with more general observations.

> Considerations like these must in the end determine the range and avenues of commerce, perhaps the fate of continents. We must develop and guide Nature's latent energies, we must utilize her inmost workshops....[7]

Thus, by 1898 not only was a potential shortage of world resources becoming recognized, but also the problem was seen as soluble only through international cooperation and scientific research.

Soddy had turned twenty-one a few days before this memorable address. He had that year received a first-class honours degree in chemistry at Oxford and remained there for two more years engaged in independent chemical research. While it cannot be confirmed on available evidence that Soddy went to Bristol for the occasion, it is certain that he received the message and made it his own.

In mid-1900 Soddy went to Canada and accepted a junior position in the chemistry department of McGill University, Montreal. Soddy collaborated there with Ernest Rutherford, who had recently been appointed one of the two Macdonald professors of physics. In a brilliant joint effort[8] Rutherford and Soddy confirmed that the phenomena of radioactivity exhibited a hitherto unknown source of energy in nature. By 1903 they established that the magnitude of this energy was more than a million times that of any previously known terrestrial source, and that it was stored within the atoms of matter. The discovery of this new source of energy was fundamental, but only a first step. Atomic disintegration accompanied by the release of this energy was still nature's secret. It proceeded at a rate which was independent of any known means of human control and seemed to be associated only with the heaviest elements. Nevertheless, Soddy was quite convinced that the atoms of all elements, not merely those like thorium and uranium, contained such a latent store of energy. There 'is associated with the internal structure of the atom an enormous store of energy which, in the majority of cases', Soddy surmised, 'remains latent and unknowable. For the heaviest elements the property of spontaneous atomic disintegration reveals its existence and enables its amount to be calculated'.[9] As understood at the time, this 'atomic energy' did not signify the conversion of mass or mass-defect into energy, but it was construed at least by Soddy as a property of atoms in general. With the experimental confirmation by Ramsay and Soddy in mid-1903 of the production of helium by radium emanation, providing a clear case of natural transmutation, the credibility of the disintegration theory of radioactivity and the associated claim of a new source of energy was greatly enhanced.

## 3. The Promise of Human Control over Atomic Energy

Soddy believed that human control over this new-found source of energy was merely a function of time and scientific ingenuity. Up to the onset of the first world war, in addition to his important scientific contributions concerning isotopes, Soddy embarked upon a public information policy. He preached the promise of artificial transmutation and the benefits that this could – and should – bring to all mankind. If atomic energy could be artificially controlled as Soddy imagined, and as he voiced in the conclusion of his 1906 address on 'The internal energy of elements', then by 'the expenditure of about one ton yearly of uranium, costing less than £1000, more energy would be derived than is supplied by all the electric supply stations of London put together. One little step, so easily anticipated in imagination . . .', said Soddy, 'divides us from this magnificent inheritance'. Furthermore, he enthused, we 'are starting the twentieth century with the prize in full view'.[10]

This prize was the hope of the future. Transmutation no longer meant what it had for the early alchemists, for the transformation of one element into another was no longer of primary interest, even if the end product were gold. The new 'gold' was energy. The fundamental problem of transmutation was now that of the controlled release of this atomic store of energy. This new problem, albeit in the first instance of concern mostly to scientists, was one that affected the whole future of civilization. It was in this vein that Soddy in 1912 enlarged upon the theme presented fourteen years earlier by Crookes on the fixation of nitrogen. His paper was appropriately entitled 'Transmutation, the vital problem of the future', and it vividly demonstrates the catholicity of his thinking in this matter even before the Great War. According to Soddy, an 'economist might not at once be prepared to accept the doctrine ... [no doubt he would today] that so far as human affairs go, wealth and available energy are synonyms, and that the poverty or affluence of this planet are primarily measured only by the dearth or abundance of the supply of energy available for its life and work...'.[11] But economists aside, even 'among scientific men the general modern method of treating energy as a fundamental commodity is often obscured by the survival of less direct and more special habits of thought. Sir William Crookes in a remarkable address to the British Association some years ago considered the great quesiton of the world's future supply of wheat...', Soddy continued. 'It happens that in this particular case we have an easily understood and specially glaring instance of what is really a universal question. For wheat is only energy' – derived in this case from the sun.[12] As for the artificial production of nitrogen compounds needed for agriculture, 'the Chilian deposits of sodium nitrate ... are certainly not inexhaustible. When they are used up one of the vital assets of agriculture will have disappeared, and thus the 'wheat problem', a phase on a small scale of the still unrecognized 'energy problem' that awaits the future, is the first to have attracted attention'.

Modern science is 'still popularly misunderstood as a creator' of the store of energy needed for its raw material, whereas it is actually 'ransacking the globe of stores which have required geological epochs of the past to lay down, and which itself it is powerless to replace. Coal must follow ultimately the threatened fate of the Chilian nitrate beds', Soddy cautioned (no doubt his admonition should also be taken to apply to gas and oil as well), 'and considering the rapidly increasing rate at which the available resources of energy are being used up, the end may come sooner than even the most pessimistic anticipate'.[13]

'As regards energy, and therefore as regards every other commodity, the modern world is undoubtedly living far beyond its income. It has recently come into a legacy from the remote past and it is living on the capital. It cannot now be very long before it wakes up to the appreciation of this fact'.[14] Yet it still took more than sixty years before the 'energy crisis' made headlines and before nations began to grasp the full social and economic implications of energy as the basic commodity. Measured against these most recent events, Soddy's comments of 1912 appear far less abstruse and of much greater portent than they did at the time.

'These stores may last out a century or two longer', Soddy continued, 'but it is obvious that the more glorious the zenith attained the more swift and sure will be the decline'.[15] Hope yet remained in the form of the controlled release of this new store of atomic energy. And the responsibility for bringing this about lay squarely on the shoulders of the scientists. As Soddy saw it before the Great War, this was their utmost duty, since the 'only possible way of escape known to science is by the solution of the problem of transmutation'.[16] Nor did he underestimate the difficulties involved. Thirty years of scientific effort lay ahead before Fermi's group in Chicago could demonstrate what Soddy could only imagine in 1912. 'The thoughtless may conclude lightly that science which has accomplished so many marvels will surely not fail when this ultimate test comes'.[17] What Soddy did not foresee at this time was the possibility that it would not be science that failed humanity in this regard, but that this ultimate achievement of science would be misused for destruction; its natural development stunted and deformed by factors beyond the control of science. To Soddy, before he became aware of the larger issue, it was the solution of the scientific problem that was most urgent, since every year 'brings appreciably nearer the inevitable coming struggle for the possession and control of the primary sources of natural energy'.[18] Without what later became known as nuclear reactors for the controlled release of this atomic energy, our civilization would be forced 'back into its former physical condition of hand to mouth existence [dependent] upon the daily supply of solar radiation'. Thus, upon the issue of transmutation 'appears to hang the whole destiny of the race, spiritual, intellectual and aesthetic as well as physical or material'.[19]

In his *Matter and Energy*, also published in 1912, Soddy showed that he had already begun to perceive the problem of transmutation within a larger perspective. The solution of this problem seemed to be the only knowledge that could 'postpone the day of reckoning' since it could 'replenish, rather than further diminish, ... a dwindling supply of energy...'.[20] But a mere postponement of the fundamental problem would, in the long run, be no solution. 'It is perfectly obvious that, with the whole planet in measurable distance of being occupied, and nations being concerned rather to preserve what they have got than to acquire more, a turning-point is being reached in the upward progress which has hitherto kept pace with the advancement of knowledge. Thoughts of economy and conservation will inevitably replace those of development and progress, and the hopes of the race will centre in the future of science'.[21]

Soddy was thirty-five when he made these observations, and the first world war was not yet visible on the horizon. When it came, the urgent scientific problem became for Soddy of secondary interest. More fundamental problems of social economics and the social responsibility of scientists began to take on greater significance for him, since unless these fundamental questions were solved in advance of a solution to the scientific problem, the achievement of transmutation could prove at best to be a mixed blessing.

## 4. The Promise is Tarnished and Achieved

The traumatic war period of 1914–1919 added a whole new dimension to the scientific problem of transmutation, and during this period Soddy sought to grasp the full meaning of the fundamental conflict and struggled in solitude to find a solution. As a scientist he recognized that the achievement of controlled atomic energy was both inevitable and a proper goal for science. Indeed it was the duty of science to forestall the otherwise inevitable 'energy problem' and to provide for the future of civilization. Yet it was becoming increasingly clear that, because of the ignorance and immaturity of society, this could prove to be a curse instead of a blessing. The chemical process recently discovered by Haber[†] had provided at least a temporary solution to the 'wheat problem'. But if the fixation of nitrogen could be used for armaments as it was during the war, and if other chemical products could be misused for the destruction of human life, then who could guarantee that the release of atomic energy would not also be perverted for warfare? Both horns of the dilemma presented a peril for civilization. Soddy therefore turned his attention from the original scientific problem to even more urgent questions of economics and society. He fervently hoped that the control of atomic energy would be *postponed* until society had become sufficiently mature to take responsibility for this achievement of scientific investigation, yet not postponed *indefinitely*, for then there would be an inevitable energy crisis. He urged his colleagues to take an interest in the moral and social implications of what they were doing. The problem was complex but clear. Why were the discoveries which science had achieved for the potential benefit of mankind in every case used for its destruction? It seemed to Soddy imperative that a solution first be found to this general problem, in order to forestall a similar fate for the achievement of transmutation.

Soddy developed this theme in a series of addresses delivered at Aberdeen throughout the war, and published in 1920 under the general title *Science and Life*. 'Imagine, if you can, what the present war would be like', he wrote in 1915, 'if such an explosive had actually been discovered instead of being still in the keeping of the future. Yet it is a discovery that conceivably might be made tomorrow, in time for its development and perfection for the use or destruction, let us say, of the next generation, and which, it is pretty certain, will be made by science sooner or later. Surely it will not need this last actual demonstration to convince the world that it is doomed, if it fools with the achievements of science as it has fooled too long in the past'.[22]

Two years later – more than sixty years ago – he warned of the dangers of atomic warfare in no uncertain terms. 'If man ever achieves this further control over Nature.... War, unless in the meantime man had found a better use for the gifts of science, would not be the lingering agony it is today. Any selected section of the world, or the whole of it if necessary, could be depopulated with a swiftness and dispatch that would leave nothing to be desired'.[23]

The period from 1919 seemed to Soddy a real race either to death and destruction or to peace and prosperity, for in that year the first successful step towards artificial control of transmutation under laboratory conditions had been announced to the scientific community.[24] Judging correctly that full control would quickly come, Soddy dragged his scientific feet and searched rather for a solution to the more fundamental problem, that every potential boon from science had in the past been utilized for the destruction instead of for the benefit of mankind. Unless this problem could be solved and the trend reversed, he forecasted the total 'wrecking' of our scientific age.[25] Soddy was dismayed that so few of his scientific colleagues were prepared to look beyond the exciting but treacherous boundary of the achievement of full human control over atomic energy. If Soddy seemed misguided to most of his colleagues, they must to him have appeared very shortsighted indeed.

Following his identification of the fundamental problem Soddy realized that there would be a tendency first to construct and utilize an atomic bomb for destruction within society, rather than an atomic reactor for the unlimited production of energy for its welfare. He saw clearly that an adequate, reliable, stable, and secure source of energy was a fundamental condition for the welfare of international society. Once atomic energy could be fully controlled and utilized as such a source, the world's total wealth could be greatly increased. There would thus no longer be any need for repeated cycles of prosperity and peace followed by famine and war.[26]

To Soddy it was a perversion of science that the achievement of transmutation should be diverted by non-scientific forces from being able to provide just this stabilizing source of energy and, instead, exploited to bolster and perpetuate a system based upon instability and inequality. Soddy considered an antiquated money system lacking a proper scientific standard of value and distribution to be the source of the difficulty. He was particularly incensed at the banking system which he felt could arbitrarily 'create' money in order to cover its debts and thereby randomly reduce the value of currency through inflation.[27]

During the next few decades, while science progressed towards the control of atomic energy, Soddy launched a blistering attack upon the evils of the incumbent monetary system. The discovery of nuclear fission in Germany in the 1930s and the rapid progress in America during the second world war is a well known story.[28] But Soddy's struggle during this same period has not yet been fully understood.

It has often been said of Soddy that he went 'off the rails' about 1919. Although he was appointed to a Chair at Oxford in that year, and received a Nobel Prize two years later, he seems thereafter to have contributed little to orthodox science and turned rather to social economics. His ensuing excommunication from the scientific community was conducted with the same righteous vigour as was his exclusion from the esoteric circle of economists. He was loved by neither and despised by all. To the one he seemed a traitor and to the other a mischievous upstart. But Soddy saw the situation differently, and, as a minority of one, took Mercutio's position of 'a plague on both your houses'.[29] If the scientific community was subservient to a decadent money system which perverted the treasures that science had provided for the benefit of society, then he wanted to be no part of that community. Neither did he endeavour to be recognized by traditional economists, for to him many of them were actually perpetuating the problem.

### 5. Soddy's Economic Views

Soddy began to take an active interest in economics and monetary reform during the early 1920s. No doubt one impetus for this was the events following Britain's formal abandonment of the gold standard in 1919. This was intended as a purely temporary manoeuvre which would permit economic readjustment and the return of Britain to the gold standard at the pre-war parity in 1925.[30] The severe recession starting in 1920 was accompanied by a large reduction in the highly inflated wages and prices over the next few years. The attempt to bring this reduction about was one of the main features of economic policy during the first half of the 1920s.[31] It was against this background that Soddy lectured and published his *Cartesian Economics* (1922), and *The Inversion of Science* (1924).[32]

The main principles of the monetary reform which he proposed had been advocated for many years by Silvio Gesell (1862–1930) in Switzerland and Germany, and by the engineer Arthur Kitson (1860–1937) in England.[33] Soddy acknowledged his indebtedness to these early reformers as well as to Major Clifford H. Douglas (1879–1952) and his school in the matter of 'credit reform and on the inefficiency, sabotage and waste that at present are inseparable from our system of distributing wealth'.[34] But Soddy stressed that he had 'been driven independently to advocate this reform after consideration of the physical nature of the defect to be redressed and its interpretation in terms of sociology and politics'.[35]

No less a reformist than John Maynard Keynes in his *General Theory* of 1936 also acknowledged his indebtedness to the 'unduly neglected prophet' Silvio Gesell,[36] and H. G. Wells pointed out that whenever the complex history of monetary reform comes to be written 'Gesell will pobably be a quite cardinal name in that story'.[37] Wells like Soddy agreed with Gesell on the 'necessity of increasing the amount of purchasing power in the world with increased production', but differed in ideology, Wells advocating socialism and Gesell 'uncontrolled individualism'.[38] Soddy took the democratic middle road. Wells thought Soddy's approach to social economics 'extraordinarily fresh and vigorous' and Soddy himself 'a pioneer'.[39] They shared a common vision of the future world economy, parting company mainly over the issue of whether this goal was being inhibited intentionally (Soddy) or unintentionally (Wells). Soddy, wrote Wells, thinks 'that the present struggle to arrest the hopeful expansion of human affairs for the sake of the gold standard is due to a conspiracy of powerful, aggressive, able men. We think that it is due to a blind convergence of fear, habit and traditional stupidity'.[40] Wells recognized that there was nothing 'to rule the plain accusation of Professor Soddy out of court',[41] but he hoped that Soddy was wrong. 'If Professor Soddy is right, . . . if it is true that the majority of able spirits among the contemporary rich are, for the sake of power and pre-eminence, deliberately impoverishing a community, which need not be impoverished, then the conception pervading this book of the progressive construction of a universally prosperous economic world community out of the current social order, is unsound'.[42]

Essential to Soddy's social economic theory was the *link* between wealth and energy and the *distinction* between wealth and money. With an adequate supply of energy there is prosperity, expansion, and development.[43] Yet 'wealth and available energy are synonyms',[44] and furthermore 'wealth is the power of purchasing',[45] so that if a stable source of energy could be provided by science, the desired goal of increasing the amount of purchasing power in the world could be achieved. The hitch was *money*, which Soddy regarded as a 'purely conventional symbol' of wealth.[46] Soddy differed 'from the reformers [Kitson, Douglas, Cass, and others], while in general agreement with their diagnosis of the industrial and economic situation, in the belief that it is not only possible for a scientific era to devise a stable monetary unit of value, but that it is of far more consequence to its social well-being, even than invariable standards of weights and measures'.[47]

In contrast to most other monetary reformers such as Douglas, Soddy preferred 'the simpler scheme of basing currency on index number'.[48] The idea of index numbers was not new. As Wells pointed out, Professor Irving Fisher (1867–1947) of Yale was already discussing money from this new point of view before the first world war.[49] In contrast to the Old Money School, which required that some commodity such as gold or silver be made the standard of value, the New Money School 'turned the matter about from the point of view of the money manipulated to consider it from the point of view of the worker, and . . . realized that money can be detached altogether from standard commodities'.[50] By the early 1930s Keynes, 'whose common-sense suggestions [were] rapidly spreading into and saturating contemporary monetary thought', was recognized as an outstanding exponent of the New Money indeas.[51] Keynes set forth very clearly the 'conception of a barometric currency based on an index number'.[52] As Soddy observed,

[It] is obviously unfair that those who depend for their livelihoods upon fixed wages, salaries, or incomes should have their purchasing power diminished, whether by the find of gold or an issue of counterfeit money. It is almost as obvious that it is unfair that the *rentier* class, whose income is derived from interest upon past debts, should share equally with the worker when ... the revenue is scientifically increased, but no new money to correspond is issued. The conclusion is thus extraordinarily simple. Average prices must be fixed by controlling the circulation of money, so that the latter is always increased or decreased proportionately with the increase or decrease of revenue. Fixing the index number by the public issue and withdrawal of currency effects this.[53]

The 'constancy of the index number would be the criterion of an honest public currency'.[54]

Wells expanded upon this point by suggesting that the final settlement to be sought

is the world organization of a New Money based on an index number, but a score of expedients may be necessary to avert disaster and gain time for the propaganda and establishment of the scientific method.... Barometric money, 'managed currency' is paper money of which the issue and withdrawal are planned to secure stability of purchasing power in general. An index price of staple commodities has to be maintained. If it falls, issue is increased; if it rises, issue is restricted. So the primary end of a sure value in wages is attained.[55]

More recently Stewart has commented that 'the monetarists think that monetary policy is very important and has been quite wrongly neglected ... while the Keynesians (to a greater extent than Keynes himself) have tended to regard monetary policy as relatively unimportant and ineffective'.[56] And yet as Lekachman notes, for a variety of reasons 'monetary policy has been enjoying a revival among economists and public officials'.[57] Even though 'Keynes himself often doubted that it would suffice to resuscitate a depressed economy, ... [there is] nothing in the Keynesian system [which] contradicts the virtues of monetary policy'.[58] The sharpest divergence of view about the importance of monetary policy, as Stewart notes, comes over the measures to be adopted to cope with inflation. 'To the Keynesians, post-war inflation has basically been a phenomenon springing from the combination of full employment on the one hand and traditional collective-bargaining and price-setting techniques on the other, and needing to be dealt with by some form of incomes policy'.[59] But to 'the monetarists inflation is, in the words of Professor Milton Friedman, "a purely monetary phenomenon"'.[60] Soddy would not have disagreed.

### 6. Scientists' Responsibility

For Soddy, the issue of the social responsibility of scientists themselves was intimately tied to the scientific question of the availability of a new source of energy and the economic question of its great potential for increasing the wealth of nations (always provided that a solution could be found to the allied problem of money). Merely to *provide* this new source of energy, as Soddy himself had preached before the first world war, was no longer sufficient. Scientists also had an obligation to participate in the vexed question of the equitable distribution of this new source of wealth. 'It is an insult to the human intelligence', wrote Soddy after the second world war, 'that the age which has solved the problem of the production of the physical requisites that enable and empower life should be unable to solve the problems of their equitable distribution, and be compelled to destroy our potential abundance periodically by internal sabotage, world wars and monetary inflation'.[61] Soddy believed that the scientific community was particularly responsible for the future of mankind. It was within their power to produce, or not to produce – to operate, or not to operate – sophisticated new products and inventions. To abstain from judgement in these matters, or to plead non-involvement, was to have made a decision by deafult to support the incumbent system. It was on this fundamental issue that Soddy, over a generation ago, broke ranks with most of his scientific colleagues.

But on this ground Soddy was not alone. He was particularly sympathetic to the views of the younger 'scientific humanists' such as Bernal and Blackett who were 'anxious to do good to suffering humanity by all the devices and possibilities of science'.[62] According to Barber, however, in the pursuit of the ideal of maximizing the potential power of science, the scientific humanists tended to absolutize science as a value in itself.[63] As socialists, they tended to identify 'capitalist economic interests' as the chief cause of 'the frustration of science'.[64] Bernal, one of the leaders of this group, wrote that

> The frustration of science is a very bitter thing. It shows itself as disease, enforced stupidity, misery, thankless toil, and premature death for the great majority, and an anxious, grasping, and futile life for the remainder. Science can change all this, but only

science working with those social forces which understand its functions and which march to the same ends.[65]

Once scientists themselves become

conscious of this frustration, the scientist is forced to inquire into the factors which control the development of science itself and to ask why it should be hampered and distorted in this way.... The demand of the scientists that science should be allowed to develop and to be used for the benefit and not for the destruction of mankind is a force ... to be reckoned with. For, unless it is met, the willing collaboration of the scientist in the present economic system will gradually be replaced by a grudging acquiescence, and finally by a blank refusal to co-operate, or a tacit sabotage.[66]

In 1935 this group of scientific humanists published a series of essays under the general title, *The Frustration of Science*,[67] and although Soddy did not share the view that socialism or communism was the solution, he did share their concern to clarify the basic problem. He therefore accepted their invitation to provide a Foreword to this volume. There is 'growing sense of social responsibility, among some individual scientific men', he wrote, but

the public expect far more from scientific men in this respect than they have as yet contributed. Individually most of them in this field are still utterly unscientific, and quite as apt as the public themselves to regard original thought on these subjects as socially dangerous and to be suppressed and those who have strayed from the path of 'pure' science in these directions as cranks or impostors.... The pioneer and bearer of a new evangel is always up against an inchoate mass, educable only when miserable and, when prosperous, too proud to learn....

Referring to the essays themselves, Soddy pointed out that here we 'read of the wilful destruction of the products and productivity of the soil, the aerial destruction of wealth ... the development of the art of spreading bacterial infection as a new war technique ... [all] elegant examples of the sort of ruling mentality now dominating the world'. It was, he thought, to a great extent up to scientists themselves to change the course of these events. 'Bitter, and justifiably so, as many of the critics of science are, surely nothing bitterer could be said of it than this, that its abundance has but enthroned the wastrel. Nor is the solution ... that science should look for a new master'. Merely to abandon capitalism for communism was no solution for Soddy.[68] As he saw it,

the solution is for the public to acknowledge its real master [viz., science], and, for its own safety, insist on being ruled ... by those who are concerned with the creation of its wealth rather than of its debts.[69] It should require that its universities and learned societies should no longer evade their responsibilities and hide under the guise of false humility as the hired servants of the world their work has made possible, but do that for which they are supported in cultured release from routine occupations, and speak the truth though the heavens fall.[70]

This was a patent and insistent invitation to the scientific community to participate in the 'open conspiracy' described a few years earlier by Wells, who referred to a 'world-wide movement conscious of itself. . .'.[71] Its hallmark: 'Tell the truth though the heavens fall, is the heroic phrasing of it'.[72]

Three decades later Bernal could claim that perhaps 'the greatest change in the position of the scientist . . . today lies in the greater self-consciousness of science and its position in society, and a greater awareness of its social function. . . . The scientist as citizen is not in the first place a scientist, only in the second'.[73] Had Soddy lived to see developments beyond the mid-1950s, he might have been encouraged. As it was, when he died in 1956 during his eightieth year, it was still too early to see just how rapidly the self-consciousness of science would accelerate. It was therefore a reproach as much as an admonition that he sounded shortly before his death: the blame for the future plight of civilization 'must rest on scientific men, equally with others, for being incapable of accepting the responsibility for the profound social upheavals which their own work primarily has brought about in human relationships'.[74]

Lewis Mumford was among those who commemorated Soddy's passing in 1956. His tribute is particularly apropos in the present context.

I met Soddy at the Sociological Society in 1920, just about the time when he was turning from his work as an experimental chemist to the problem of economics, and particularly to the problem of money, as a medium for making energy available for human life. This preoccupation with economic problems doubtless seemed an aberration to many other people besides his fellow-scientists. Without doubt it involved a sacrifice of Soddy's special gifts in the field of Chemistry; but it showed an alertness to the social responsibilities of the scientific mind that was exemplary; so that if there had been more Soddys in the world, with the same alertness and the same sensitivity, our age would have been prepared – as even now it is not – to cope with the vast problems . . . which nuclear energy has raised. Soddy demonstrated, by his own too lonely example, the necessary transcendence of the limits of specialization which must become general, if our achievements are not to be undone by the crippled minds who know their isolated field not wisely but too well, and who know life itself, in all its manifold possibilities and challenges, not at all. By his example, yes even by his 'failure', he has become one of the heralds of a new age, whose coming awaits a whole generation of Soddys.[75]

## 7. Conclusion

One of the great misconceptions of society, according to Barber, has been to think 'that science is wholly a-moral'. Scientific humanism points out that science 'rests on a definite set of moral values intimately related to the values of "liberal" society as a whole'.[76] While not sharing the socialist bent of the scientific humanists, Soddy was clearly disabused of this misconception about value-free science quite early on. 'Weighty enough influences still prevent any approach to the realization of Utopia, *but*', he wrote in 1919, 'they are not now physical but moral'.[77] For him it was only a matter of time before the scientific question of transmutation would be solved. Since energy is wealth, the prosperity of the world was assured, at least in principle. He therefore turned his attention from the relatively simple scientific problem to the more complex and urgent moral, economic, and social ones. Soddy was indeed concerned with a variety of interrelated issues. That the integrity and catholicity of his thought eluded most of his contemporaries, and was taken instead to be the aberration of an otherwise astute mind, is perhaps but one indication of just how far he was ahead of his time. That the significance of what he had to say is all too evident today, is perhaps an indication of just how far the 'open conspiricy' has now advanced.

*RR4 – Box 639*
*Colborne, Ontario*
*KOK 150 Canada*

## Notes

* Reprinted from *The British Journal for the History of Science* **12**: 42 (November 1979): 261–76.

† The Haber Nitrogen Fixation Process is still the dominant method for fixation of nitrogen and ultimately for the production of fertilizers and nitric acid. Without it, munitions production would have been severely hampered during both world wars.

1 Frederick Soddy, *Matter and Energy*, London, 1912, p. 251. F. Soddy, 'Transmutation, the Vital problem of the Future', *Scientia*, 1912, **11**, 199.

2 F. Soddy, 'Advances in the Study of Radio-Active Bodies', two lectures to the Royal Institution on 15 May and 18 May 1915, as recorded in *The Royal Institution Friday*

*Evening Lectures, 1907–1918*, privately bound at the Royal Institution, London, n.d. The original MS is in the Bodleian Library, Soddy-Howorth Colection, 58 (Alton, 111). The lectures are apparently unpublished but were reviewed in *Engineering*, 1915, **99**, 604. The quotation is from the MS, p. II, 9.

[3] F. Soddy, *Science and Life: Aberdeen Addresses*, London, 1920, p. 6. The original address entitled 'Science and Life' was given in December 1918.

[4] F. Soddy, concluding statement from Typescript-A of Soddy's biographical records, in the Bodleian Library, Soddy-Howorth Collection, 4 (Alton, 2).

[5] F. A. Paneth, 'Classical Radioactivity and its Sequence', *Nature*, 1950, **166**, 800.

[6] Sir William Crookes, 'Presidential Address', *Report of the British Association for the Advancement of Science*, 1898, p. 18.

[7] *Ibid.*, p. 19.

[8] T. J. Trenn, *The Self-Splitting Atom: A History of the Rutherford-Soddy Collaboration*, London, 1977.

[9] F. Soddy, 'Radio-Activity', *The Electrician*, 1904, **52**, 646.

[10] F. Soddy, 'The Internal Energy of Elements', *Journal of Proceedings of the Institution of Electrical Engineers*, 1906, **37**, 8. The cost of production is now over ten times greater; cf. M. Hansen, 'Trends in Uranium Supply', *International Atomic Energy Agency Bulletin*, 1976, **18**, 16–27.

[11] Soddy, 'Transmutation', *op. cit.* (1), p. 187.

[12] *Ibid.*, p. 198.

[13] *Ibid.*, pp. 199–200.

[14] *Ibid.*, p. 200.

[15] *Ibid.*

[16] *Ibid.*

[17] *Ibid.*

[18] *Ibid.*, p. 201.

[19] *Ibid.*, p. 202.

[20] Soddy, *Energy*, *op. cit.* (1), p. 246.

[21] *Ibid.*, pp. 246–7.

[22] F. Soddy, *op. cit.* (3), p. 36. The original address, 'Physical Force – Man's Servant or His Master?', was given in November 1915. The quotation is from the section dealing with science and war. See also Bernard Barber, *Science and the Social Order*, New York, 1970 (reprint of 1952 edn.), p. 288.

[23] Soddy, *op. cit.* (3), p. 107. The original address 'The Evolution of Matter' was given in February 1917.

[24] T. J. Trenn, 'The Justification of Transmutation: Speculations of Ramsay and Experiments of Rutherford', *Ambix*, 1974, **21**, 53–77.

[25] F. Soddy, *The Wrecking of a Scientific Age*, London, 1927.

[26] F. Soddy, *Dishonest Money; or Why a Larger Pay-Packet Now Buys Less Than It Did*, London, 1950.

[27] F. Soddy, *Wealth, Virtual Wealth and Debt: The Solution of the Economic Paradox*, London, 1926, *The Arch-Enemy of Economic Freedom: What Banking Is, What First It Was, and Again Should Be*, London, 1943.

[28] E. N. Hiebert, *The Impact of Atomic Energy*, Newton, Kansas, 1961; and Alice K. Smith, *A Peril and a Hope*, Chicago, 1965.

[29] F. Soddy, *Money Reform as a Preliminary to All Reform*, Birmingham, 1950, p. 2.
[30] M. Stewart, *Keynes and After*, 2nd edn., Harmondsworth, 1972, p. 60.
[31] *Ibid.*
[32] F. Soddy, *Cartesian Economics: The Bearing of Physical Science upon State Stewardship*, London, 1922; *The Inversion of Science: And a Scheme of Scientific Reformation*, London, 1924.
[33] Silvio Gesell, *Die natürliche Wirtschaftsordnung durch Freiland und Freigeld*, 3rd edn., Arnstadt & Thüringen, 1919; *Die Verwirklichung des Rechtes auf den vollen Arbeitsertrag durch die Geld- und Bodenreform*, Hauts-Geneveys, Switzerland, 1906. Arthur Kitson, *A Scientific Solution of the Money Question*, London, 1894; *The Money Problem*, London, 1903; *A Fraudulent Standard; An Exposure of the Fraudulent Character of Our Monetary Standard, with Suggestions for the Establishment of an Invariable Unit of Value*, London, 1917.
[34] Soddy, *Inversion*, *op. cit.* (32), pp. 5–6; Clifford Hugh Douglas, *Credit-Power and Democracy; with a Draft-Scheme for the Mining Industry*, London, 1920; *Economic Democracy*, London, 1920.
[35] Soddy, *Inversion*, *op. cit.* (32), p. 5.
[36] Stewart, *op. cit.* (30), p. 76; John Maynard Keynes, *The General Theory of Employment, Interest and Money*, London, 1936.
[37] H. G. Wells, *The Work, Wealth and Happiness of Mankind*, London, 1932, p. 365.
[38] *Ibid.*
[39] *Ibid.*, p. 487.
[40] *Ibid.*, p. 492.
[41] *Ibid.*, p. 488.
[42] *Ibid.*, p. 492.
[43] Soddy, *Energy*, *op. cit.* (1), p. 251.
[44] Soddy, 'Transmutation', *op. cit.* (1), p. 187.
[45] Soddy, *op. cit.* (3), p. 27.
[46] *Ibid.*; cf. F. Soddy, *Money Versus Man: A Statement of the World Problem from the Standpoint of the New Economics*, London, 1931; *The Role of Money*, London, 1935.
[47] F. Soddy, 'Unemployment and Hope', *Nature*, 1930, **125**, 346; cf. W. G. Linn Cass, 'Unemployment and Hope', *Nature*, 1930, **125**, 225.
[48] Soddy, *Inversion*, *op. cit.* (32), p. 6.
[49] Wells, *op. cit.* (37), pp. 349, 365. Irving Fisher, *The Making of Index Numbers*, New York, 1922; *The Nature of Capital and Income*, New York, 1906; *The Purchasing Power of Money*, New York, 1911. Cf. Friedrich A. Hayek, *The Pure Theory of Capital*, London, 1941, *passim*.
[50] Wells, *op. cit.* (37), p. 365.
[51] *Ibid.*
[52] *Ibid.*; J. M. Keynes, *A Tract on Monetary Reform*, London, 1923, 'constituted an attack on one of the main pillars of the British economic system', and, according to Stewart, *op. cit.* (30), p. 22, Keynes found himself 'virtually alone' in his rejection of Britain's return to the pre-war gold standard system; cf. Keynes, *A Treatise on Money*, London, 1930.
[53] Soddy, *Inversion*, *op. cit.* (32), p. 23.
[54] *Ibid.*, p. 41.

[55] Wells, *op. cit.* (37), p. 371.
[56] Stewart, *op. cit.* (30), pp. 228–9.
[57] Robert Lekachman, *The Age of Keynes: A Biographical Study*, Harmondsworth, 1966, p. 88.
[58] *Ibid.*
[59] Stewart, *op. cit.* (30), p. 229.
[60] *Ibid.*
[61] Soddy, *op. cit.* (26), p. 20.
[62] Barber, *op. cit.* (22), p. 292.
[63] *Ibid.*
[64] *Ibid.*, pp. 292–3.
[65] John Desmond Bernal, *The Social Function of Science*, Cambridge, Mass., 1967 (original edn., London, 1939), p. xv.
[66] *Ibid.*, p. 386.
[67] Sir Daniel Hall *et al.*, *The Frustration of Science*, London, 1935 (reprinted New York, 1975). This contains a Foreword by Soddy on the inversion of science; Chapter I by Sir Daniel Hall on science and agriculture; Chapter II by J. G. Crowther on aviation; Chapter III by J. D. Bernal on science and industry; Chapter IV by V. H. Mottram on medicine; Chapter V by E. Charles on the invention of sterility with respect to the population explosion; Chapter VI by P. A. Gorer on bacterial warfare; and Chapter VII by P. M. S. Blackett entitled 'The Frustration of Science'. Blackett insists that science must become involved in social and economic issues: 'Unless society can use science, it must turn anti-scientific, and that means giving up the hope of the progress that is possible. This is the way that capitalism is now taking, and it leads to Fascism' (p. 139). The *only* other alternative envisioned here is socialism, which encourages science, instead of rejecting science like fascism. These claims of Blackett clearly set the tone of the volume.
[68] Soddy, *op. cit.* (67), pp. 7–9. Soddy refused to accept the socialist gambit and followed neither the rank individualism of Gesell nor the creed of Wells. He preferred democracy. 'The dispute between Socialism and individualism centres on one point, the question of the ownership of revenue-producing enterprise.... [The alternative], democracy stands for a fair deal between the individual and the community.... Do not trust labels. Look well to realities. An equally sinister fate is as likely to come from Socialism itself [as from extreme individualism or Fascism]. If it does not understand the origins of social injustice, it is as likely to end in an oligarchy, supported by military power...'; Soddy, *Inversion*, *op. cit.* (32), p. 47). Cf. F. Soddy, 'What I Think About Socialism', *Socialist Review*, August 1928.

A decade later, Soddy expounded further on this point in the light of the political polarity that had hardened in the meantime. 'There is growing up among the masses, who have no part in or liking for the power-politics of their rulers, a feeling of fatalism that the threatened conflict is inevitable. According to the Marxist doctrine, capitalism carries within itself the seeds of its own collapse, so that they are coming to regard communism as the only alternative. My thesis is that the threatened collapse of our Western civilization has nothing to do with the political issues between capitalism and communism, but is the consequence of its false money system. Not only the impending collapse but its whole development, culminating in this very concentration of power into two opposing world-groups, has been due to the creation and issue of money having passed out of the hands of the several nations, at first into the hands of private bankers and now into the supranational Money Power which has usurped sovereignty over the whole Western World. The alternative is for the West certainly not communism, which . . . would be the substitution of one form of tyranny for another . . . [but] the restoration to the several nations of their traditional sovereign powers over the creation of money, and . . . an honest money system, with money of constant purchasing power, operated, not for profit, but . . . as a public service, like those provided for the maintenance of just weights and measures. . .'; Soddy, *op. cit.* (29), p. 2.

[69] Soddy considered science to be the creator of wealth, but under the inversion of science the 'exploiters of the wealth of the world are not its creators'. Under this inversion, from 'the point of view of the community, capital is not wealth but debt, the not owning by the community of the resources of the planet whereon it resides. . .'; Soddy, *op. cit.* (3), pp. 23–4. The rise of the 'scientific civilization' which Soddy advocated would meet with many obstacles, as he acknowledged just after the first world war. 'The war being now over, it is not out of place to add that an even greater danger than neglect awaits the scientific investigator, the danger that he along with every other creative element in the community will be remorselessly shackled and exploited to bolster up the present discredited social system. There is abundant evidence since the war that science rules the world, and he who would aspire to rule it must first rule science'; Soddy, *op. cit.* (3), p. 109; cf. F. Soddy, 'Social Relations of Science', *Nature*, 1928, **141**, 784–5.

[70] Soddy, *op. cit.* (67), p. 9.

[71] Wells, *op. cit.* (37), p. 781.

[72] *Ibid.*, p. 777.

[73] Bernal, *op. cit.* (65), pp. xxxiv–xxxv. These comments appear in his essay 'After Twenty-Five Years' included in this volume and originally published in M. Goldsmith and A. Mackay (eds.), *Society and Science*, London, 1964.

[74] Soddy, *op. cit.* (4).

[75] Lewis Mumford, 'Commemorative Message', in *Commemoration of Professor Frederick Soddy*, London, 1956/8, pp. 11–12. The lack of alertness and sensitivity here described is represented typically by Soddy's biographer and close colleague in earlier research on isotopes, Sir Alexander Fleck: 'The basis of his thought for the whole of his life after coming to Oxford was an endeavour to resolve the weaknesses of our modern civilized life . . . [Soddy's] main conclusion was that "it was entirely due to the fictitious money system which arose contemporaneously with the birth of the scientific civilization and that now was being purposefully and consciously used to frustrate it and to preserve the earlier civilization founded on slavery". . . . To most of us,' Fleck admitted, 'these writings have little attraction and the best that can be said of them is that they represented an attempt to base a monetary system on an assessment of energy quantities. . . . He continued to write and think on these lines until the 1950s but the impact of these writings tended to diminish'; Alexander Fleck, 'Frederick Soddy', *Biographical Memoirs of Fellows of the Royal Society*, 1957, **3**, 210–11. Twenty years later it can be asked whether Soddy's views should not have attracted more attention within the scientific community, and whether to have ignored them can be justified.

[76] Barber, *op. cit.* (22), p. 293.

[77] Soddy, *op. cit.* (3), p. 172.

HERMAN E. DALY

# 11. THE ECONOMIC THOUGHT OF FREDERICK SODDY*

Almost always the men who achieve these fundamental inventions of a new paradigm have been either very young or very new to the field whose paradigm they change. – Thomas S. Kuhn. *The Structure of Scientific Revolutions* (1962), p. 89.

## 1. Introduction

Frederick Soddy (1877–1956) is best known as a pioneering chemist who collaborated with Rutherford in studying radioactive disintegration, predicted the existence of and coined the name for isotopes, and was a major contributor to the modern theory of atomic structure. For these achievements he was elected a Fellow of the Royal Society in 1910 and was awarded the Nobel Prize in 1921. He was a member of the Swedish, Italian, and Russian academies of science. During his career he held university positions at McGill, Glasgow, Aberdeen, and from 1919 onward, Oxford (Fleck, 1957).

Although an enthusiastic believer in scientific progress and in the possibility of a society in which the fruits of scientific knowledge would be shared by all, Soddy was acutely aware that history supported the view that science has proved as much a curse as a blessing to humanity. Nor could he accept the comfortable view that scientists have no responsibility for the uses to which their work is put. Even though others (bankers and economists) bore, in his view, a far greater burden of guilt for the misuse of knowledge, scientists could not plead innocent. The world's real problem was faulty economics, not faulty chemistry, and for the second half of his nearly eighty years economics replaced chemistry as the center of his intellectual life.

Soddy realized earlier than most the theoretical possibility of atomic energy. Since his own work had contributed to the discovery that this vast energy potential existed, it was natural for him to ask, 'what sort of a world it would be if atomic energy ever became available' (*Wealth*, p. 28). His answer (written in 1926) was clear:

*George B. Kauffman (ed.), Frederick Soddy (1877–1956)*, 199–218.

If the discovery were made tomorrow, there is not a nation that would not throw itself heart and soul into the task of applying it to war, just as they are now doing in the case of the newly developed chemical weapons of poison-gas warfare.... If [atomic energy] were to come under existing economic conditions, it would mean the *reductio ad absurdum* of scientific civilization, a swift annihilation instead of a none too lingering collapse [*Wealth*, p. 28].

For Soddy, the problem was to change economic conditions in order eventually to make the world safe for atomic energy and other fruits of science. There must be something radically wrong with economic thought and institutions in order for the gift of scientific knowledge to have become such a threat. Soddy was thus led to a radical critique of economics.

It is interesting that Soddy's concern about the destructive potential of atomic energy was considered extreme at the time. Another Nobel laureate, Robert A. Millikan, commented:

... since Mr. Soddy raised the hobgoblin of dangerous quantities of available subatomic energy [science] has brought to light good evidence that this particular hobgoblin – like most of the bugaboos that crowd in on the mind of ignorance – was a myth.... The new evidence born of further scientific study is to the effect that it is highly improbable that there is any appreciable amount of available subatomic energy for man to tap [Millikan, p. 121].

Millikan, of course, turned out to be wrong, but the underlying faith that he went on to express is still held by many, namely that one may 'sleep in peace with the consciousness that the Creator has put some foolproof elements into his handiwork, and that man is powerless to do it any titanic physical damage' (*ibid.*). As R. L. Sinsheimer recently noted, 'Scientific endeavor rests upon the faith that our scientific probing and our technological ventures will not displace some key element of our protective environment and thereby collapse our ecological niche' (Sinsheimer, p. 24). It now seems evident that the only protective element the Creator put into his handiwork is man's capacity for moral insight and restraint, which is far from foolproof. With the benefit of hindsight we can see that Soddy was the true prophet and that the scientific establishment, represented by Millikan, was whistling in the dark.[1] Far from believing in providential 'foolproof elements' built into creation, Soddy was convinced that the economic system contained built-in elements for assuring the destruction of creation, once science gave man the power. The key problem therefore was to discover and correct the errors in our economic thinking and institutions, a task which Soddy tackled with both moral fervor and the systematic logic of an experienced scientist.

Perhaps the most intriguing thing about Soddy the economist is that he started his inquiry with a mind both highly intelligent and completely free from the preconceived paradigm of the orthodox economists, for whom he had an undisguised contempt. The contempt was mutual. With the significant exception of Frank Knight, to be discussed later, Soddy's work was ignored by economists. Unlike the American positional astronomer, Simon Newcomb, who also came to economics from the physical sciences, Soddy came as a critic, not a student, and remained an outsider. Newcomb liked economics, did not believe that his pre-World War I America was in mortal danger from an increasingly powerful but misdirected application of science, and wrote a fairly orthodox *Principles of Political Economy* (1885) which demonstrated that he had done his economics homework, and had earned the right to try to make economics just a bit more scientific. Soddy, on the other hand, considered economics a pseudoscience in need of a totally new beginning. John Ruskin, not Ricardo, Mill, or Marshall, was his inspiration.

The not surprising consequence of this approach was that Soddy was and continues to be written off as a crank. In fact, Soddy's economics seems to have been something of an embarrassment to everyone but Soddy. The *Times Literary Supplement* (p. 565), in reviewing his major economic work (*Wealth, Virtual Wealth, and Debt*) remarked that it was sad to see a respected chemist ruin his reputation by writing on a subject about which he was quite ignorant. Nor had the verdict changed thirty years and several books later, when in 1956 an obituary in *Science* lamented that, 'Some . . . knew him only as . . . a "crank" on the subject of monetary policy. . . . His fanatical devotion to schemes of this sort, derided by the orthodox economists,. . . was surprising to many who knew him first as a pioneer in chemical science' (Russell, p. 1069). This neglect of Soddy's economics is unfortunate because, although Soddy is admittedly unconvincing in his frequent attribution of war and all other evils to fractional reserve banking, he nevertheless has much to teach us, and in fact anticipated the recent contribution of N. Georgescu-Roegen (1971) in providing economics with a partial foundation in thermodynamics, the physics of usefulness.

The fact that Soddy might have learned more from economists than he did does not mean that economists have nothing to learn from Soddy. The approach here taken is to think of him somewhat as an intelligence from Mars which looked at economic issues in a different way, and to try sympathetically to understand him and render him intelligible to modern economists. In what follows I attempt to summarize and explain Soddy's critique of economics.

## 2. The Neglected Physical Basis of Economics

Soddy's basic philosophical approach to economics might be called materialism without reductionism. We must recognize the fundamental dualism of the material and the spiritual and resist 'monistic obsessions' (*Cartesian Economics*, p. 6). Economics occupies the middle ground between matter and spirit, between the electron and the soul:

> In each direction possibilities of further knowledge extend *ad infinitum*, but in each direction diametrically away from and not towards the problems of life. It is in this middle field that economics lies, unaffected whether by the ultimate philosophy of the electron or the soul, and concerned rather with the *interaction*, with the middle world of life of these two end worlds of physics and mind in their commonest everyday aspects, matter and energy on the one hand, obeying the laws of mathematical probability or chance as exhibited in the inanimate universe, and, on the other, with the guidance, direction and willing of these blind forces and processes to predetermined ends [*ibid.*].

Soddy rejects the monism of 'Ultra-Materialism':

> I cannot conceive of inanimate mechanism, obeying the laws of probability, by any continued series of successive steps developing the powers of choice and reproduction any more than I can envisage any increase in the complexity of an engine resulting in the production of the 'engine-driver' and the power of its reproducing itself. I shall be told that this is a pontifical expression of personal opinion. Unfortunately, however, for this argument, inanimate mechanism happens to be my special study rather than that of the biologist. It is the invariable characteristic of all shallow and pretentious philosophy to seek the explanation of insoluble problems in some other field than that of which the philosopher has first hand aquaintance [*ibid.*, p. 7].

Yet a proper materialism must be one of the foundation stones of economics. In fact, 'without phosphorus no thought' (*Story of Atomic Energy*, p. 129) is an axiom that all philosophers and ethicists should be required to memorize. What mechanical science teaches economics is that

> life derives the whole of its physical energy or power, not from anything self-contained in living matter, and still less from an external deity, but solely from the inanimate world. It is dependent for all the necessities of its physical continuance primarily upon the principles

of the steam-engine. The principles and ethics of human law and convention must not run counter to those of thermodynamics [*Cartesian Economics*, p. 9].

The last sentence is very significant because it provides the basis for many of Soddy's criticisms of the economy as a presumed perpetual motion machine. For men, like other heat engines, the physical problems of life are energy problems. Pre-nineteenth-century man lived on energy revenue (sunlight captured by plants, the 'original capitalists'). Present-day man augments this revenue by consuming energy capital (coal, the 'stored sunlight of palaeozoic summers'). While man can use fuel-fed machinery to lighten labor, he can feed his internal fires only with new sunshine, or rather the energy of new sunshine as transformed through the good offices of the plant. Life thus depends on a continuous flow of energy, and hence the enabling requisites of life must partake of the nature of a flow rather than only a stock. There are limits to the degree that this flow can be stored for future use. A significant part of the requisites of life must come to us as a current flow or 'revenue' that cannot in any physical sense be converted to a stock and indefinitely stored for later use. Like the manna which God sent to the Hebrews in the wilderness, the revenue is renewed daily, must be gathered in amounts sufficient for the day (neither too much nor too little), and breeds worms and becomes foul if accumulated too much in excess of current needs (*Exodus* 16: 17–20). Stocks of assets, to the extent that we can maintain them against the ravages of entropy, are aids and accessories in improving our ability to tap the energy revenue, but the revenue itself cannot be significantly increased, and it cannot be saved except to a limited degree. Indeed, the very maintenance of our accumulated stock of physical wealth against the destructive force of entropy requires the renewing power of the low-entropy 'revenue' flow. True, nature has stored energy in coal, but it took geologic epochs of time, and we are only able to unstore it. Furthermore the 'flamboyant period' of using up the capital stock of coal was perceived by Soddy as a 'very passing phase', after which the constraints imposed by living on energy revenue would be more clearly seen and unmistakably felt.

For Soddy the basic economic question was 'How does man live?' and the answer was 'By sunshine'. The rules that man must obey in living on sunshine, whether current or palaeozoic, are the first and second laws of thermodynamics. This in a nutshell is 'the bearing of physical science upon state stewardship'. Wealth is for Soddy 'the humanly useful forms of matter and energy' (*The Arch Enemy*, p. 6). Wealth has both a physical dimension, matter-energy subject to the laws of inanimate mechanism, and a teleological dimension of usefulness, subject to the purposes imposed by mind and will. Soddy's concept of wealth reflects his fundamental dualism and his belief that the middle world of life and wealth is concerned with the interaction of the two end worlds of physics and mind in their commonest everyday aspects. That Soddy concentrated on the physical dimension in order to repair the consequences of its past neglect should not be allowed to lead one to suppose that he proposed a monistic physical theory of wealth, a misinterpretation which, we will see, was fostered by Frank Knight.

## 3. The Major Confusion: Wealth Versus Debt

The fundamental error of economics is the confusion of wealth, a magnitude with an irreducible physical dimension, with debt, a purely mathematical or imaginary quantity. The positive physical quantity, two pigs, represents wealth and can be seen and touched. But minus two pigs, debt, is an imaginary magnitude with no physical dimension:

> Debts are subject to the laws of mathematics rather than physics. Unlike wealth, which is subject to the laws of thermodynamics, debts do not rot with old age and are not consumed in the process of living. On the contrary, they grow at so much per cent per annum, by the well-known mathematical laws of simple and compound interest.... For sufficient reason, the process of compound interest is physically impossible, though the process of compound decrement is physically common enough. Because the former leads with the passage of time ever more and more rapidly to infinity, which, like minus one, is not a physical but a mathematical quantity, whereas the latter leads always more slowly towards zero, which is, as we have seen, the lower limit of physical quantities [*Wealth*, p. 70].

The ruling passion of the age is to convert wealth into debt in order to derive a permanent future income from it – to convert wealth that perishes into debt that endures, debt that does not rot, costs nothing to maintain, and brings in perennial interest (*Money Versus Man*, p. 25). No individual could amass the physical requirements sufficient for maintenance during his old age, for like manna it would rot. Therefore he must convert his non-storable surplus into a lien on future revenue, by letting others consume and invest his surplus now in exchange for the right to share in the increased future revenue. The revenue is 'a river of perishable and consumable wealth, steadily flowing to waste whether consumed by human beings or by rats and worms' (*Inversion of Science*, p. 24). But since future annual revenue is limited, there is a corresponding limit on the extent to which present surpluses can be exchanged for perennial streams of future revenue. Soddy emphasizes that the present surplus accumulation can never be changed into future revenue in any physical sense, but only *ex*changed for it under social conventions. Although it may comfort the lender to think that his wealth still exists somewhere in the form of 'capital', it has been or is being used up by the borrower either in consumption or investment, and no more than food or fuel can it be used again later. Rather it has become debt, an indent on future revenues to be generated by future sunshine. 'Capital', says Soddy, 'merely means unearned income divided by the rate of interest and multiplied by 100' (*Cartesian Economics*, p. 27).

Although debt can follow the law of compound interest, the real energy revenue from future sunshine, the real future income against which the debt is a lien, cannot grow at compound interest for long. When converted into debt, however, real wealth 'discards its corruptible body to take on an incorruptible' (*Money Versus Man*, p. 28). In so doing, it appears 'to afford a means of dodging Nature' (p. 24), of evading the second law of thermodynamics, the law of random, ravage, rust, and rot. The idea that people can live off the interest of their mutual indebtedness (*Wealth*, p. 89) is just another perpetual motion scheme – a vulgar delusion on a grand scale. Soddy seems to be saying that what is obviously impossible for the community – for everyone to live on interest – should also be forbidden to individuals, as a principle of fairness. If it is not forbidden, or at least limited in some way, then at some point the growing liens of debt holders on the limited revenue will become greater than the future producers of that revenue will be willing or able to support, and conflict will result. The conflict takes the form of debt repudiation. Debt grows at compound interest and as a purely mathematical quantity encounters no limits to slow it down. Wealth grows for a while at compound interest, but, having a physical dimension, its growth sooner or later encounters limits. Debt can endure forever; wealth cannot, because its physical dimension is subject to the destructive force of entropy. Since wealth cannot continually grow as fast as debt, the one-to-one relation between the two will at some point be broken – i.e. there must be some repudiation or cancellation of debt. The positive feedback of compound interest must be offset by counteracting forces of debt repudiation, such as inflation, bankruptcy, or confiscatory taxation, all of which breed violence. Conventional wisdom considers the latter processes pathological, but accepts compound interest as normal. Logic demands, however, that either we constrain compound interest in some way, or accept as normal and necessary one or more of the counteracting mechanisms of debt repudition.[2] As Soddy put it,

> You cannot permanently pit an absurd human convention, such as the spontaneous increment of debt [compound interest], against the natural law of the spontaneous decrement of wealth [entropy] [*Cartesian Economics*, p. 30].

The perpetual motion delusion of living on debt has arisen in the following way, Soddy says:

> Because formerly ownership of land – which, with the sunshine that falls on it, provides a revenue of wealth – secured, in the form of rent, a share in the annual harvest without labor or service, upon which a cultured and leisured class could permanently establish itself, the age seems to have conceived the preposterous notion that money, which can buy land, must therefore itself have the same revenue-producing power [*Wealth*, p. 106].

If debt and money are the units of measure by which we account for and keep track of the production and distribution of physical wealth, then surely the units of measure and the reality being measured cannot be governed by different laws. Soddy's 'acid test is that no monetary accountancy be allowed that could not be done equally well by physical counters' (*The Arch Enemy*, p. 24). If wealth cannot grow at compound interest for long, then debt should not either. If wealth cannot be created *ex nihilo* then how can we allow money (debt) to be created *ex nihilo* (and just as easily destroyed)? Worse, how can we tolerate the fact that money is both created *ex nihilo* and lent at compound interest, while at the same time serving as a unit of measure for wealth which is incapable of either of those 'conjuror's tricks'? This brings us to money, the topic which most occupied Soddy's attention.

## 4. The Monetary Flaw

The main defect in the economic system was, for Soddy, the practice of fractional reserve banking whereby the private banking system was enabled to create money, thus appropriating what he called the Virtual Wealth of the community, which it then lent at interest. The concept of 'virtual wealth' plays a key role in Soddy's analysis. Essentially it is the aggregate value of real wealth which individuals in the community voluntarily abstain from holding in order to hold money instead. In order to escape the inconvenience of barter everyone must hold money, which could be exchanged for real wealth, but is not. In Soddy's words, 'This aggregate of exchangeable goods and services which the community continuously and permanently goes without (though *individual* money owners can instantly demand and obtain it from other individuals) the author terms the Virtual Wealth of the community' (*Role of Money*, p. 36).

If everyone tried to exchange his money holdings for real assets it could not be done, because all real assets are already owned by someone, and in the final analysis someone has to end up holding the money. So Virtual Wealth does not really exist as actual wealth over and above the value of real assets, which is why it is called Virtual. Yet people behave as if Virtual Wealth were real, because at an individual level money is easily exchangeable for physical assets. The phenomenon of Virtual Wealth must occur in a monetary economy, unless the money is itself a commodity that circulates at its commodity value.

The value of each unit of money, or the inverse of the 'price-index', is simply the Virtual Wealth divided by the total aggregate of money held.

Soddy gives the following summary of the nature and importance of Virtual Wealth:

> Money is now a form of national debt, owned by the individual and owed by the community, exchangeable on demand for wealth by transference to another individual. Its value or purchasing power is not directly determined by any positive or existing quantity of wealth, but by the negative quantity or deficit of wealth, the ownership and enjoyment of which is voluntarily abstained from without the payment of interest, by the owners of the money, to suit their individual business and domestic affairs and convenience. The aggregate of this deficit is called the Virtual Wealth of the community, and it measures the value of all the money owned by the community, which is forced by the necessity of exchanging its produce to act as though it possessed this amount of wealth more than it actually does possess. The Virtual Wealth of a community is not a physical but an imaginary negative wealth quantity. It does not obey the laws of conservation, but is of psychological origin [*Wealth*, p. 295].

Virtual Wealth varies with the size of population and national income and the business and payment habits of the community. It is only when Virtual Wealth is constant that we can equate the value of a unit of money to the ratio of Virtual Wealth to aggregate money held. Soddy believed that Virtual Wealth, though not constant, was far less variable than the money supply.

Who benefits from Virtual Wealth? In a sense the whole community does, since it is the price of avoiding barter, or more precisely the price of avoiding the waste of a full commodity currency which uses costly resources (gold) to perform a function that could be performed by paper or by abstract accounting units. In another sense, Virtual Wealth is like seigniorage, the difference between the monetary value and the commodity value (cost of production) of the money token. With the advent of credit money the commodity value of the token becomes nil and seigniorage or Virtual Wealth is the full monetary value of the money issued – or rather the equivalent in foregone utility. The analogy with seigniorage suggests a further answer to the question of who benefits from Virtual Wealth. It is the issuer of fiat money, whoever first puts it in circulation, that gets the seigniorage. The ancient prerogative of the crown has been usurped, not by the modern State, the crown's legitimate heir, but by the private banking system, which 'has corrupted the purpose of money from that of an exchange medium to that of an interest-bearing debt' (*Wealth*, p. 296). Moreover the very existence of the bulk of our money depends upon this debt never being liquidated. The very existence of money now becomes a source of private income, and the total money supply becomes a 'concertina' expanding to fuel a boom, and contracting with debt repayment and default thereby reinforcing a slump.

Soddy's concept of Virtual Wealth bears an interesting relation to the modern debate about whether fiat money is a part of the net wealth of the community. Pesek and Saving (1967) argue that it is, whereas others, such as James Tobin (1965), argue that it is not. Soddy says that fiat money is Virtual Wealth. Individuals voluntarily hold money balances rather than an equivalent value in real assets in order to escape the enormous inconvenience of barter. Virtual Wealth is the utility cost of holding money. The fact that the benefits are worth more than the costs does not make the costs disappear, and does not convert money into wealth. The social institution of money may be regarded as a form of collective patrimony in the same sense as an efficient legal code or an advanced technology. But the money commodity itself need not be, and in the case of fiat money is not, a productive asset. Indeed, the very advantage of fiat money is to free resources from being tied up in money so that more real assets may be produced with the resources. We count the extra real assets made possible by fiat money as a part of the aggregate wealth of the community, but not the paper chits themselves.

Soddy's notion of Virtual Wealth is actually very close to what James Tobin terms the 'fiduciary issue':

> The community's wealth now has two components: the real goods accumulated through past real investment and fiduciary or paper 'goods' manufactured by the government from thin air. Of course, the nonhuman wealth of such a nation 'really' consists only of its tangible capital. But, as viewed by the inhabitants of the nation individually, wealth exceeds the tangible capital stock by the size of what we might term the fiduciary issue. This is an illusion, but only one of the many fallacies of composition which are basic to any economy or society. The illusion can be maintained unimpaired as long as the society does not actually try to convert all of its paper wealth into goods [Tobin, p. 676].

For Soddy, banks do not really make loans, because a loan implies that the lender gives up what the borrower receives. When a bank lends money it gives up nothing, creating the deposits *ex nihilo* up to the limit set by reserve requirements.[3] The real 'lender' is the community at large whose money balances lose in purchasing power with the issue of new money. We know the new money will be spent and increase demand, because the borrower who gets it would not pay interest just to increase his idle balances. Prices are bid up since *ex nihilo* creation of money (demand) can increase much more rapidly than can the *ex materia* creation of new physical wealth (supply). But the more direct line of causation is simply that relatively constant Virtual Wealth divided by more pounds means each pound is worth less. Money should not bear interest as a condition of its existence, but only when genuinely lent by an owner who gives it up to a borrower. Banks are like counterfeiters who lend false money, accept their own false money in repayment and destroy it, but receive the interest in real money transferred to them by the rest of the community, and which is not destroyed. Banks create and destroy money with no understanding of the 'laws that correlate its quantity with the national income' (*Wealth*, p. 296). Also by continually changing the value of money as they create and destroy it, the banking system converts the pound sterling into a rubber yardstick, in effect making a mockery of all physical measurement standards, since 'yards per pound' or 'gallons per pound' become variable magnitudes, even though yards and gallons be fixed.

At first sight it may seem odd that one who analyzes the economy with the concepts of physical science should focus so much on money, instead of real resources, matter, energy, and the rest. But, of course, it is precisely the fact that money seems to have escaped the laws of conservation and entropy that led Soddy to conclude that the flaw in the system must lie with the 'conjuror's tricks' of modern bankers, who

> have been allowed to regard themselves as the owners of the virtual wealth which the community does *not* possess, and to lend it and charge interest upon the loan as though it really existed and they possessed it. The wealth so acquired by the impecunious borrower is not given up by the lenders, who receive interest on the loan but give up nothing, but is given up by the whole community, who suffer in consequence the loss through a general reduction in the purchasing power of money [*Wealth*, p. 296].

A further contradiction arises from the interest-bearing national debt being used as collateral security by bondholders who borrow from banks. Banks create a deposit (new money) for the borrowing bondholder and charge him interest. The public is taxed to enable the government to pay interest on the bond to the bondholder who, in effect, passes the interest on to the bank. Soddy draws the conclusion that 'taxes are thus paid to the bank for doing what the taxes were imposed to prevent being done, namely, the increase of the currency. Otherwise, there would have been no reason for the State to borrow at interest if it had not wished to prevent the increase of the currency' (*Wealth*, pp. 195, 298). Soddy considers this the final *reductio ad absurdum* of the monetary system.

## 5. Reform Measures

Three basic reforms are suggested by Soddy to restore honesty and accuracy to the economic system: a 100% reserve requirement for banks; a policy of maintaining a constant price-index; and freely fluctuating exchange rates internationally.

With a 100% reserve requirement banks could no longer create money, and that basic function, along with the seigniorage prerogative, or the ownership of Virtual Wealth, would be restored to the State, which would again become the sole 'utterer' of money. Banks would have to exist by charging for their legitimate services, i.e. those that do not require the creation of money.

What principle is to govern the State in issuing money? Money is to be created or destroyed by the State as needed in order to keep the purchasing power of money constant. A price index will be devised by the National Statistical Authority. If the index has a tendency to fall over time, the government will finance its own activities by printing money.

Alternatively it might lower taxes, or use the newly created money to redeem interest-bearing national debt. In other words deflation would be corrected by some form of money-creating government deficit. If the index shows a rising tendency the government will raise taxes (or issue interest-bearing national debt) and not spend the revenue. Inflation would be corrected by a money-destroying government surplus. Soddy makes an analogy between the price index and the governor on a steam engine. Both provide a mechanism of stabilizing feedback. The then existing system suffered from destabilizing feedback, since the money supply would expand during a boom and contract during a slump, thereby reinforcing the original tendency.

Equilibrium in balance of payments with the rest of the world would be achieved by freely fluctuating exchange rates which would tend to establish a kind of purchasing power parity among currencies. International flows of gold and the consequent inflationary and deflationary pressures on national economies would thereby be eliminated, thus easing the task of keeping the internal purchasing power of the currency constant. Furthermore, the need for tariffs and other interferences with free trade designed to correct international payments imbalances, major causes of international conflict, would have been eliminated.

Soddy's proposals have nothing in common with those of Silvio Gesell or Major Douglas, or other famous 'monetary cranks'. Soddy respected these men for raising important questions, but concluded that in their proposals for reform they were just as guilty of appealing to 'conjuror's tricks' as were the orthodox money men. Far from advocating 'funny money schemes', Soddy considered the existing canons of sound finance to be elaborate mystifications obscuring the most blatant 'funny money' practices carried on in the interest of the bankers and their class, to the detriment of society. These socially dishonest though perfectly legal practices, along with the attempt to convert wealth into debt internationally and live off the interest received from other countries, plus the waning of the 'flamboyant period' of energy capital consumption, of which 'imperialism marks its final bid for survival' (*Cartesian Economics*, p. 12), would lead inexorably to international conflict and to the misuse of the gifts of science in warfare.

Reform of both economic understanding and the economic system in the light of physical and moral first principles is the *sine qua non* of a civilization capable of using knowledge for good rather than evil. 'Let us have an end of the pretence that economics should not be concerned with morals' (*Role of Money*, p. 214). As a minimum morality, economics must surely insist on a system of honest weights and measures underlying exchange; yet the current monetary system with its fluctuations in purchasing power subverts honest measure and gives a false accounting of the physical realities underlying the production and distribution of wealth.

## 6. The Relevance of Soddy's Economic Thought Today

Soddy's insistence that the first and second laws of thermodynamics must be the starting point of economics (*Role of Money*, pp. 4, 5) is a fundamental insight the relevance of which has grown as we have come to discover that neither the sources of low entropy inputs nor the sinks for high entropy waste outputs are infinite. Probably the most important economic treatise of the last forty years is Nicholas Georgescu-Roegen's *The Entropy Law and the Economic Process* (1971), which demonstrates that the economic process is entropic in its physical coordinates; that wealth is an open system, a structure maintained in the midst of a throughput that begins with the depletion of low entropy matter-energy and ends with the return of an equal quantity of polluting high entropy matter-energy back to the environment; that in contrast to the reversibility of mechanical phenomena, entropic phenomena are characterized by irreversibility, a fatal weakness of the mechanistic epistemology of standard economics; and that there is a critical asymmetry between our two sources of low entropy. The last point refers to the fact that solar low entropy (Soddy's revenue) is nearly infinite in total amount but strictly limited in its rate of flow to earth, whereas terrestrial low entropy (concentrated minerals in the earth's crust) is strictly limited in total amount, but can be used up at a rate of our own choosing. Economic development since the industrial revolution has been in the direction of ever less reliance on the abundant solar flow and towards dependence on the relatively scarce terrestrial stock. This is what Soddy called the 'flamboyant period', destined to be short-lived.

Evidently Georgescu-Roegen was unaware of the writings of Soddy on this subject, because he never cites Soddy. No one is more scrupulously honest and painstaking in citing the work of others than Georgescu-Roegen, and this omission is pointed out only to indicate the extent of Soddy's obscurity as an economist. Similar comments apply to

Kenneth Boulding,[4] who has also related economics to thermodynamics, without mentioning Soddy, and to the present author as well. This omission is understandable because after all Soddy was a chemist not an economist, and his economic writings all bore titles indicating only the monetary nature of his economic work, or such uninformative titles as *Cartesian Economics*. Only the subtitle of the latter, 'The Bearing of Physical Science Upon State Stewardship', gives any hint of the nature of his most important and original contribution to economics. But the fact remains that Soddy anticipated the basic insights of Georgescu-Roegen and Boulding regarding the relation of economics and thermodynamics, and deserves recognition as a pioneer in a line of thinking which I believe will one day be dominant.

Soddy was also a pioneer in recognizing the moral responsibility of science, and in realizing ahead of others that new knowledge, while it might not be permanently 'forbidden', can certainly be 'inopportune' under existing social and moral conditions, even to the extent of being lethal to the civilization that made it possible (Sinsheimer, p. 24).

Was Soddy successful in his effort to disover the flaws in the economic system that corrupted the fruits of science and led to war? Would 100% reserves, a constant price index, and flexible exchange rates make the world safe for atomic energy? Is it true that whether science emancipates or destroys humanity depends on a 'minor technical point in a banking system', as Soddy claimed (*Inversion of Science*, p. iv)? One may reasonably doubt it. In fact it seems that at this point Soddy himself was 'seeking the solution to insoluble problems in some field other than that of which the philosopher has first-hand acquaintance' – to recall his own jibe at the mechanistic biologists. But the fact that Soddy exaggerated the efficacy of his suggested reforms does not mean that his analysis is unimportant. Neither the specific proposals nor the reasoning underlying them can be fairly dismissed as those of an outsider or a monetary crank who just does not understand economics.[5] Flexible exchange rates have come into being already, and Soddy was arguing their virtues at a time when most economists were wedded to the gold standard. The new humility born of the theoretical anomaly of simultaneous inflation and unemployment and the demonstrated inability of orthodox 'monetary cranks' to deal with persistent inflation could conceivably lead to a reconsideration of the constant price-index and 100% reserve requirements. Of course some of these policies have had other champions besides Soddy, some with very respectable academic credentials, such as Henry Simons and Irving Fisher (see Simons 1948 and Fisher 1935).

It is curious that Irving Fisher never mentions Soddy in his writings on 100% money. Soddy, however, in a pamphlet written in 1943 refers to Fisher: 'Some years later, after the great depression in the U.S.A., an American economist, Professor Irving Fisher of Yale University, put forward a scheme which in its original form was practically identical [to Soddy's 'pound for pound banking' plan] and which he termed 100% money' (*The Arch Enemy*, p. 11). Soddy's plan was published in 1926, Fisher's in 1935. Soddy seems to regard the near identity of plans as an interesting and encouraging coincidence and in no way suggests that Fisher had copied or even been influenced by him.

Although a great enthusiast for science and technology, Soddy could not share the popular obsession with unlimited growth. Even if continual economic growth were possible, it would at some point become senseless. On this point Soddy quotes John Ruskin, whom he greatly admired as an economist: 'Capital which produces nothing but capital is only root producing root; bulb issuing in bulb, never in tulip; seed issuing in seed never in bread. The Political Economy of Europe has hitherto devoted itself to the multiplication . . . of bulbs. It never saw or conceived such a thing as a tulip' (*Money Versus Man*, p. v).

Soddy held that 'economic sufficiency is the essential foundation of all national greatness and progress' (*Money Versus Man*, p. 12). But sufficiency means 'enough' and growth beyond 'enough' is just 'seed issuing in need never in bread'. Soddy does not define 'sufficiency', but it is clear that any definition must respect the limits of energy revenue from the sun as captured by plants, and the entropic limits on the possibility of storing up wealth for future use, as well as the teleological constraint implied by a defined (i.e. limited) purpose from which low entropy matter-energy derives its value dimension, and thus becomes wealth. Not all matter-energy is capable of becoming wealth; only low entropy matter-energy has the physical potential for usefulness, for receiving the imprint of information and purpose. Since the entropy law says in effect that potential gets used up, then scarcity must increase over the long run. But what bothered Soddy was not the scarcity implications of entropy, since he believed that science could more than offset increasing scarcity for a very long time yet. The truly scarce factor for Soddy was not low entropy, but our ability to keep from blowing up scientific civilization with the increasing power that science made available. We persist in applying those powers toward the impossible goal of making the real world of matter-energy conform to the purely mathematical law of compound interest. This leads to debt cancellation, conflict, and war. Orthodox growth economics notwithstanding, the best evidence is that the earth is not growing at all, much less at a rate equal to the rate of interest. The attempt to pit an absurd human convention against a natural law is not only foolish, but highly dangerous.

The absurdity of infinite growth has been the most carefully ignored anomaly in the paradigm of modern economics. As Soddy put it,

> If Christ, whose views on the folly of laying up treasures on earth are well known, had put by a pound at this rate, it should now be worth an Octillion, and Tariff Reform would be of little help to provide that, even if you colonized the entire stellar universe. . . . It is this absurdity which inverts society, turns good into evil and makes orthodox economics the laughing stock of science. If the consequences were not the familiar atmosphere of our daily lives they would be deemed beyond the legitimate bounds of the most extravagant comic opera [*Inversion of Science*, p. 17].

A contemporary unsympathetic reviewer, economist A. G. Silverman, was at least forthright enough to face the issue and to attempt a reply:

> In criticism of the above theory, it may be asked how can the receivers of interest, if they live on this income, take advantage of the law of compound interest; and if they reinvest this 'unearned' income why cannot the law of compound interest approximately hold for physical capital as well as debt [Silverman, p. 277]?

The first question is sensible but irrelevant because Soddy never suggested that one could take advantage of compound interest without reinvesting at least part of the interest income. The second part of the qustion, however, reveals that the questioner had no conception of the difference between physical and purely mathematical quantities, and must have made Soddy despair of ever communicating with economists. Perhaps by 'approximately' Professor Silverman meant 'for a limited time period'. But then we must ask what happens at the end of that limited time period, and how long it is.

Probably the most favorable review that Soddy got from an economist came from none other than Frank Knight, who began by confessing that

> Somewhat to the reviewer's surprise this book [*Wealth, Virtual Wealth, and Debt*] has proven well worth the time and effort of a careful reading. Surprising because, in general, when the specialist in natural sciences takes time off to come over and straighten out the theory of economics he shows himself even dumber than the academic economist, and

because, in particular, Soddy's pamphlet on *Cartesian Economics* which we read some years ago did not promise to set a new precedent in this regard [Knight, p. 732].

Knight went so far as to call the book (p. 732) 'brilliantly written and brilliantly suggestive and stimulating'. I would like to be able to appeal to the authority of Frank Knight to support my own favorable evaluation of Soddy's economics, but unfortunately our particular appreciations of Soddy conflict. Knight considers Soddy's practical theses concerning money to be 'highly significant and theoretically correct' (*ibid.*), a judgment with which I do not basically disagree, but consider a bit too kind, since Soddy certainly exaggerated the significance of his practical theses, however correct they may be. Concerning the physical basis of economics and the relation to thermodynamics, however, Knight is very negative (*ibid.*):

His effort to establish a conception of physical wealth, subject to a principle of conservation and interpretable in relation to physical energy, must be briefly dismissed.

Knight's grounds for dismissal are opaque:

Magnitudes of wealth and productive capacity ... change absolutely whenever a human being changes his (or her!) mind; and the mass-energy relations of mind-changes are as unimportant in this connection as they are obscure – if their very existence is anything but a metaphysical inference based on the monistic bias of the scientific intellect [Knight, p. 732].

Whatever that may mean, it is surely odd that anyone who read *Cartesian Economics* (recall the first two quotations in Section 3 above) could even obliquely accuse Soddy of 'monistic bias'. Furthermore Soddy had no theory of conservation of the *value* dimension of wealth (which may change with mental states), but only insisted that the physical dimension of wealth is subject to the laws of thermodynamics regardless of mind changes or financial conventions, and that this fact is not trivial. If the fact that magnitudes of wealth and productive capacity change absolutely whenever a human being changes his mind were the whole truth, then how easy it would be to make everyone wealthy – all we would need do to double wealth would be to change our minds! Then wealth could grow as fast as debt, since it would be free from its physical body. It would be quite unfair to accuse Knight of such simple-minded angelism, but it strikes me as equally unfair of Knight to treat Soddy as a simple-minded physical reductionist. It is true that Soddy emphasized the physical aspect of wealth in order to correct for its neglect by economists, a procedure which, if Knight's attitude is representative, Soddy was certainly justified in adopting.

In view of the fact that Soddy's critique of money stemmed directly from his prior physical analysis, it is strange that Knight could so categorically reject the latter while enthusiastically embracing the former, although it is conceivable that one could arrive at the right monetary conclusions for the wrong physical reasons. Knight offers the following support for Soddy's views on money:

> In the abstract, it is absurd and monstrous for society to pay the commercial banking system 'interest' for multiplying severalfold the quantity of medium of exchange when (a) a public agency could do it at negligible cost, (b) there is no sense in having it done at all, since the effect is simply to raise the price level, and (c) important evils result, notably the frightful instability of the whole economic system [*ibid.*].

Knight deserves much credit for having been the only reputable economist to have taken Soddy seriously, even though in my opinion he missed Soddy's main contribution. But then so did everyone else until now, when, in the light both of Georgescu-Roegen's masterly reuniting of economics with its physical base and of the current recognition of the critical importance of energy, the prior contribution of Soddy has become visible enough for anyone to see. Soddy was in many ways fifty years ahead of his time.

The author is grateful for support from the Rockefeller Brothers Fund during the period in which this article was written. Helpful comments on an earlier draft were received from T. Beard, W. Campbell, E. Cook, G. Hardin, S. Farber, G. Smith, and an anonymous referee. The author, of course, is solely responsible for all contents of the article.

*Louisiana State University*
*Baton Rouge, LA 70803*
*U.S.A.*

## Notes

* Reprinted from *History of Political Economy* **12**, 4. (1980, Duke University Press)

1 In fairness to Millikan it should be noted that in concluding his vigorous defense of science he did temper his optimism with the following caution: 'I am not in general disturbed by expanding knowledge or increasing power, but I begin to be disturbed when this comes coincidentally with a decrease in the sense of moral values. If these two occur

together, whether they bear any relationship or not, there is real cause for alarm' (Millikan, p. 129).
[2] This point has been forcefully made by biologist Garrett Hardin. See his (with Carl Bajema) *Biology: Its Principles and Implications*, 3d ed. (San Francisco, 1978), p. 257.
[3] When a bank lends to A it forgoes the opportunity of making the same loan to B, so in that sense there is an opportunity cost in allocating the virtual wealth among borrowers, but there is no opportunity cost to the bank in acquiring the Virtual Wealth in the first place.
[4] In fact, Boulding told me he was very much aware of Soddy the scientist, having slept through his chemistry lectures at Oxford, but knew nothing of his economic writings. As for sleeping through chemistry lectures, even the writer of one obituary tribute remarked that it would be idle to pretend that Soddy was a successful classroom teacher.
[5] For such a dismissal see A. G. Silverman (1972).

## Bibliography

Fisher, I.: 1935, *100% Money*, New York.

Fleck, A.: 1957, 'Frederick Soddy', *Biographical Memoirs of Fellows of the Royal Society* **3**, 203–16.

Georgescu-Roegen, N.: 1971, *The Entropy Law and the Economic Process*, Cambridge, Mass.

Knight, F. H.: 1927, *Review of Wealth, Virtual Wealth and Debt. Saturday Review of Literature*, p. 732.

Millikan, R. A.: 1930, 'Alleged Sins of Science', *Scribner's Magazine* **872**, 119–30.

Pesek, B. P. and Saving, T. R.: 1967, *Money, Wealth, and Economic Theory*, New York.

Russell, A. S.: 1956, 'F. Soddy, Interpreter of Atomic Structure', *Science*, pp. 1069–70.

Silverman, A. G.: 1927, *Review of Wealth, Virtual Wealth, and Debt, American Economic Review*, pp. 275–78.

Simons, H.: 1948, *Economic Policy for a Free Society*, Chicago.

Sinsheimer, Robert L.: 1978, 'The Presumptions of Science', *Daedalus*, pp. 23–35.

Soddy, F.: 1920, *Science and Life*, London.

Soddy, F.: 1922, *Cartesian Economics*, London.

Soddy, F.: 1924, *The Inversion of Science*, London.

Soddy, F.: 1926, 1933, *Wealth, Virtual Wealth, and Debt*, London.

Soddy, F.: 1933, *Money Versus Man*, New York.

Soddy, F.: 1934, *The Role of Money*, London.

Soddy, F.: 1943, *The Arch Enemy of Economic Freedom* (pamphlet), Oxford.

Soddy, F.: 1949, *The Story of Atomic Energy*, London.

*Times Literary Supplement*, London, 26 Aug. 1926, p. 565.

Tobin, J.: Oct. 1965, 'Money and Economic Growth', *Econometrica* **33**.

# BIOGRAPHICAL NOTES ON CONTRIBUTORS

Dr. LAWRENCE BADASH
*Department of History*
*University of California, Santa Barbara*
*Santa Barbara, CA 93106*
*U.S.A.*

Lawrence Badash, born in Brooklyn, N.Y. on May 8, 1934, received his B.S. in physics from Rensselaer Polytechnic Institute (1956) and his Ph.D. in the history of science from Yale University (1964), where he was Instructor (1964–65) and Research Associate (1965–66). He is now Professor of the History of Science at the University of California, Santa Barbara, where he has taught since 1966. He has been a consultant for the American Chemical Society (1961), Time-Life, Inc. (1963), and the General Research Corporation (1968), a NATO Fellow (1965–66), and a Guggenheim Fellow (1984–85). A recipient of research grants from the National Science Foundation, the University of California's Institute on Global Conflict and Cooperation, and the American Philosophical Society, he has been a member of the Advisory Board of the Rand Corporation Reactor History Project (1975–77) and of the University of California Institute of Appropriate Technology (1977–79). In 1983 he was director of the University of California's Summer Seminar on Global Security and Arms Control. Since 1981 he has been a member of the California Workshop on American Foreign and Defense Policy. He has spent extended periods abroad, especially at Cambridge and Paris. His interests lie in the history of modern physics, especially radioactivity, nuclear physics, and the development of military and civilian uses of atomic energy. His books include *Rutherford and Boltwood: Letters on Radioactivity* (1969), *The Rutherford Correspondence Catalog* (1974), *Radioactivity in America: Growth and Decay of a Science* (1979), *Reminiscences of Los Alamos, 1943–1945* (1980), and *Rutherford, Kapitza, and The Kremlin* (1985).

*George B. Kauffman (ed.), Frederick Soddy (1877–1956)*, 219–226.

Dr. A. D. CRUICKSHANK
*Nuclear Engineering International*
*Quadrant House*
*Sutton Surrey*
*SM2 5AS*
*England*

A. D. Cruickshank took a first degree in chemistry at Oxford University, where he went on to pursue postgraduate research, first on the life and work of Frederick Soddy and later on the inter-war relationships between science, industry, and government in the United Kingdom. In 1981 he completed a doctoral thesis on innovation in new energy technologies with the Science Policy Research Unit of the University of Sussex. The author of articles on the history of science and on energy policy, Dr. Cruickshank carried out research at SPRU on energy research and development policy in the European Communities before joining the magazine *Nuclear Engineering International*, of which he is now Deputy Editor.

Dr. HERMAN E. DALY
*Department of Economics*
*Louisiana State University*
*Baton Rouge, LA 70803*
*U.S.A.*

Herman E. Daly is Alumni Professor of Economics at Louisiana State University and holds a B.A. from Rice University and a Ph.D. from Vanderbilt University. He has served as a Ford Foundation Visiting Professor at the University of Ceará, Brazil, as a Research Associate at Yale University, as a Visiting Fellow at the Australian National University, and as a Senior Fulbright Lecturer in Brazil. He has been a member of the Committee on Mineral Resources and the Environment of the National Academy of Sciences, the Nominating Committee of the American Economic Association, and the recipient of Louisiana State University's Distinguished Research Master Award. He has served on the boards of advisors of numerous environmental organizations, including Friends of the Earth. His interest in economic development, resources, environment, and population has resulted in some fifty articles in professional journals and anthologies as well as four books, including *Steady-State Economics* (1977) and *Economics, Ecology, Ethics* (1980). During the past thirteen years he has given some 130 invited lectures at universities and conferences both in the U.S. and abroad. On various occasions he has served as a consultant to the Ford Foundation, the World Bank, AID, World Council of Churches, and the Joint Economic Committee of Congress.

Dr. NORMAN FEATHER†
*Department of Physics*
*University of Edinburgh*
*Maxwell Building*
*Edinburgh EH9 3JZ*
*Scotland*

Norman Feather, F. R. S., born on Nov. 16, 1904, received his B.A. and B.Sc. degrees from Cambridge University and the University of London, respectively, in 1926 and his Ph.D. from Cambridge University in 1930. The author of numerous books and articles, he was Leverhulme Fellow and Lecturer in Physics at Liverpool University (1935–36), Lecturer in Physics, Cambridge University (1936–45), and Fellow, Trinity College, Cambridge (1936–45). A Fellow and Past President of the Royal Society of Edinburgh and a Fellow of the Cambridge Philosophical Society, Dr. Feather died on Aug. 14, 1978.

MICHAEL I. FREEDMAN
*P.O. Box 30459*
*Baltimore, Maryland 21270*
*U.S.A.*

Michael I. Freedman received his bachelor's degree in chemistry from Duke University, and subsequently studied the history of science in the graduate program at The Johns Hopkins University.

Dr. GEORGE B. KAUFFMAN
*Department of Chemistry*
*California State University, Fresno*
*Fresno, CA 93740*
*U.S.A.*

George B. Kauffman, Professor of Chemistry at California State University, Fresno, is the author or editor of twelve books and more than 550 technical papers, reviews, and encyclopedia articles on chemistry, chemical education, and the history of science. A former Guggenheim Fellow (1972), Dr. Kauffman is the recipient of many awards and honors, among them the Dexter Award in the History of Chemistry (1978), the Outstanding Professor Award of the California State University and Colleges System (1973), the Manufacturing Chemists Association Award for Excellence in College Chemistry Teaching (1976), the Lev Aleksandrovich Chugaev Jubilee Diploma and Bronze Medal of the U.S.S.R. Academy of Sciences (1976), and the California State University and Colleges System Exceptional Merit Service Award (1984). He has been Contributing Editor of the *Journal of College Science Teaching* since 1973, Editor of the American Chemical Society's History of Chemistry Audiotape series (1975–81), Contributing Editor of the *Hexagon* since 1980, and Contributing Editor ('Historical Sketches' feature) of *Polyhedron* since 1982. In 1982 he was awarded grants from the National Endowment for the Humanities and Svenska Institutet (The Swedish Institute) for his research project 'A Humanist Genius as Amateur Scientist: August Strindberg's Chemical and Alchemical Studies and Their Influence on His Literary and Dramatic Productions'.

Dr. ALEKSANDR N. KRIVOMAZOV
*Institute for the History of Science and Technology*
*Academy of Sciences of the U.S.S.R.*
*Staropanskii per. 1/5*
*Moscow K-12*
*U.S.S.R.*

Aleksandr Nikolaevich Krivomazov, born on May 30, 1947 at Alma-Ata, Kazakh S.S.R., studied physics at the Moscow Engineering Physical Institute. Upon his graduation in 1971, he entered the Institute for the History of Science and Technology, Academy of Sciences of the U.S.S.R., to pursue further study and scientific work in the history of science under Academician B. M. Kedrov. In 1975 he obtained his candidate physics-mathematics science degree with a thesis on the history of the discovery of the Soddy-Fajans displacement law. He is the author or coauthor of five books and more than fifty articles.

Dr. GLENN T. SEABORG
*Lawrence Berkeley Laboratory*
*Berkeley, California 94720*

Dr. Glenn T. Seaborg, who shared the 1951 Nobel Prize in Chemistry with Edwin M. McMillan, originated the actinide concept for placing the heaviest elements in the periodic system and is the co-discoverer of elements 94 through 102 and 106. Active in governmental and university affairs, he was Chairman of the U.S. Atomic Energy Commission (1961–71), Associate Director of the Lawrence Berkeley Laboratory (1954–61, 1971-present), and Chancellor of the University of California (1958–61). He has been President of the American Association for the Advancement of Science (1972) and the American Chemical Society (1976), has been elected an honorary member of many scientific societies, both domestic and foreign, and is the recipient of numerous honors and awards.

Dr. S. B. SINCLAIR
*Department of Humanities*
*Teesside Polytechnic*
*Middlesbrough, Cleveland TS1 3BA*
*England*

After graduating with a degree in chemistry from Imperial College, London, Dr. Sinclair turned to a study of the early history of radioactivity for his doctoral thesis. He is now Senior Lecturer in History and Philosophy of Science and Technology, Humanities Department of

Teesside Polytechnic where he lectures on the history of ideas for students in the humanities and on historical studies in science, technology, and society for scientists and engineers. He has directed research projects on the relationships between scientists and the military during World War I and on early scientific societies.

Dr. JAN W. VAN SPRONSEN
*Veenendaalkade 463*
*2547 AL The Hague*
*The Netherlands*

Johannes Willem van Spronsen, born in The Hague on Feb. 7, 1928, studied chemistry at the Technische Hogeschool, Delft (1945–46) and the University of Leiden (1946–52). Since 1954 he has been a teacher at the Municipal Gymnasium, Alkmaar and since 1960 Assistant Professor of Educational Chemistry at the University of Utrecht. His doctoral dissertation, '100 Years' History of the Periodic System of the Elements', was published in 1969 as the first of his four books. A popular lecturer in the Netherlands and other European countries, he was Visiting Professor at the University of Ghent (1971). His honors include the Silver Medal of the Société des Sciences, Arts et Lettres (1969), the J. B. Gillis Award of the Royal Flemish Academy of Sciences (1970), the Dexter Award in the History of Chemisty (1975), and the corresponding membership of Deutsches Gesellschaft für Geschichte der Medizin, Naturwissenschaften und Technik (1977). For 25 years he has been chairman of the Commission for the History of the Royal Dutch Chemical Society and member of the board of the Chemical Magazine (Netherlands).

Dr. MINORU TANAKA†
*Group of the History of Science*
*Tokyo Institute of Technology*
*Ookayama, Tokyo 152*
*Japan*

†Minoru Tanaka (1907–1978) graduated from Tokyo Imperial University in 1930 and was immediately appointed Assistant at the Tokyo Institute of Technology. Six years later he resigned and began to study the history of science. After World War II he was appointed Lecturer at the Tokyo Institute of Technology in 1946 and promoted to Professor in 1959. He gave lectures on the history of science and obtained his Ph.D. in the history of science in 1961. He retired in 1968 and moved to a private university. He wrote more than thirty books on the history of science and attended the International Congresses on the History of Science held in Warsaw (1965), Paris (1968), and Edinburgh (1977). For an éloge see 'Minoru Tanaka (1907–1978). In Memoriam.' *Japn. Stud. Hist. Sci.*, **17** (1978): 158 (in English), and Kazuo Mafune, *Kagakusi Kenkyu* (*J. Hist. Sci. Japan*), **18** (1979), Ser. No. 129: 53–55, and by Toshio Yamazaki, *ibid.*, 55–57 (in Japanese).

Dr. THADDEUS J. TRENN
*RR4 – Box 639*
*Colborne, Ontario*
*KOK 150 Canada*

Thaddeus J. Trenn, born in Chicago, Illinois in 1937, received master's degrees in physics and philosophy from Notre Dame University and worked for two years as a government physicist. Beginning in 1964 he lectured on philosophy for several years in California and Wisconsin. At the University of Wisconsin he pursued a doctoral program in the history of science, spending alternate academic years at Oxford and Cambridge. Awarded a postdoctoral research fellowship by the Max Planck Society, two years later he was appointed to the Department of History of Science at the University of Regensburg, West Germany. The author of several books, Dr. Trenn has been active in interdisciplinary teaching and research on the historical, philosophical, political, and cultural aspects of science at various universities, government agencies, and research organizations both in Europe and America.

Dr. KAZUO YAMASAKI
*1-28 Zendana-cho, Mizuho*
*Nagoya 467*
*Japan*

Kazuo Yamasaki was born in 1911, graduated from Tokyo Imperial University in 1933, and worked under Yuji Shibata during 1933–41 as a graduate student and research assistant. In 1941 he moved to the newly established Nagoya University and taught inorganic chemistry and spectrochemical analysis as Assistant Professor (1941–43) and Professor (1943–74). He is now Professor Emeritus at Nagoya University. His fields of research include coordination chemistry and archaeological chemistry (chemical analyses of ancient Chinese and Japanese pigments, bronzes, and glasses), and he is the auhor or editor of thirty books and more than 180 technical papers in these fields. In 1959 he received the award of the Japan Academy for the 'Study of the Mural Painting of the Daigoji Pagoda', a joint work with art historians.

# INDEX OF NAMES

# INDEX OF SUBJECTS